AF581696

Nanowire Field-Effect Transistor (FET)

Nanowire Field-Effect Transistor (FET)

Special Issue Editors

Antonio García-Loureiro
Karol Kalna
Natalia Seoane

MDPI • Basel • Beijing • Wuhan • Barcelona • Belgrade • Manchester • Tokyo • Cluj • Tianjin

Special Issue Editors
Antonio García-Loureiro
Universidade de Santiago de Compostela
Spain

Karol Kalna
Bay Campus Swansea University
UK

Natalia Seoane
Universidade de Santiago de Compostela
Spain

Editorial Office
MDPI
St. Alban-Anlage 66
4052 Basel, Switzerland

This is a reprint of articles from the Special Issue published online in the open access journal *Materials* (ISSN 1996-1944) (available at: https://www.mdpi.com/journal/materials/special_issues/Nanowire_Field_Effect_Transistor).

For citation purposes, cite each article independently as indicated on the article page online and as indicated below:

LastName, A.A.; LastName, B.B.; LastName, C.C. Article Title. *Journal Name* **Year**, *Article Number*, Page Range.

ISBN 978-3-03936-208-0 (Hbk)
ISBN 978-3-03936-209-7 (PDF)

Contents

About the Special Issue Editors

Antonio García-Loureiro (associate professor) received a Ph.D. degree from the University of Santiago de Compostela, Santiago de Compostela, Spain, in 1999. He is an associate professor at the Department of Electronics and Computer Science, University of Santiago de Compostela. He was a previously a post-doctoral visiting researcher at the University of Edinburgh (2001) and the University of Glasgow (2004). His current research interests are multidimensional simulations of nanoscale transistors and solar cells. He has 91 peer-reviewed papers, more than 250 publications, and over 20 invited talks.

Karol Kalna received an M.Sc. (Hons.) in solid state physics and a Ph.D. in condensed matter from Comenius University, Bratislava, Czechoslovakia/Slovakia, in 1990 and 1998, respectively. He has been a research scientist at the Institute of Electrical Engineering, Slovak Academy of Sciences, and a postdoctoral researcher at the University of Glasgow, U.K. He is currently a professor of electronics and the leader of the Nanoelectronic Devices Computational Group, Swansea University, Wales, U.K. He held an EPSRC Advanced Research Fellowship from 2007 to 2012, and has been pioneering III-V MOSFETs for digital applications since 2002. He has 100 peer-reviewed papers, more than 250 publications, and over 20 invited talks.

Natalia Seoane (Ramon y Cajal Research Fellow) received a Ph.D. degree from the University of Santiago de Compostela, Santiago, Spain, in 2007. She was a visiting post-doctoral researcher at the University of Glasgow, Glasgow, U.K., from 2007 to 2009 and at Edinburgh University, Edinburgh, U.K., in 2011. She was a Marie Curie IEF Research Fellow at Swansea University, Swansea, U.K., from 2013 to 2015. She is currently based at the University of Santiago de Compostela. She has 45 peer-reviewed papers and more than 100 publications.

Editorial

Special Issue: Nanowire Field-Effect Transistor (FET)

Natalia Seoane [1,*], Antonio García-Loureiro [1,*] and Karol Kalna [2,*]

1 Centro Singular de Investigación en Tecnoloxías Intelixentes, University of Santiago de Compostela, 15782 Santiago de Compostela, Spain
2 Nanoelectronic Devices Computational Group, College of Engineering, Swansea University, Swansea SA1 8EN, Wales, UK
* Correspondence: natalia.seoane@usc.es (N.S.); antonio.garcia.loureiro@usc.es (A.G.-L.); k.kalna@swansea.ac.uk (K.K.)

Received: 3 April 2020; Accepted: 14 April 2020; Published: 14 April 2020

Abstract: This Special Issue looks at recent developments in the research field of Nanowire Field-Effect Transistors (NW-FETs), covering different aspects of technology, physics, and modelling of these nanoscale devices. In this summary, we present seven outstanding articles on NW-FETs by providing a brief overview of the articles' content.

Keywords: nanowire field-effect transistors; metal gate; material properties; fabrication; modelling; variability

---

In the last years, the leading semiconductor chip manufacturing companies have introduced multi-gate, non-planar transistors into their core business with digital and analogue applications to memories, processors, and radio-frequency (RF) communication in order to achieve a larger integration on chip, increase their speed and thus data throughput and, most importantly, to reduce energy consumption. There is intense research underway to keep developing these multi-gate transistors and overcome their limitations in order to continue a transistor scaling while to further improve performance and to reduce energy consumption.

Nanowire field-effect transistors (NW-FETs) are nowadays one of the strongest contenders to replace fin field-effect transistors (FinFETs) in the following semiconductor technological nodes, because of their superior electrostatic control of the channel transport via the gate around their entire channel. This Special Issue looks at recent developments in the research field of NW-FETs. For this reason, the articles include different aspects of the physics, technology, and modelling of nanoscale NW-FETs. We present seven outstanding articles on NW-FETs by providing a brief summary of the articles' content.

The article by Yoon et al. [1] reports on the influence of the gate and drain voltages on the charge transport properties in a zinc oxide NW-FET through temperature and voltage-dependent measurements. They found that variable-range hopping charge transport dominates the conduction in the zinc oxide NW-FET in the low temperature regime of 4 K to 100 K, whereas the thermal activation charge transport is dominant from 150 K to 300 K, diminishing the space charge-limited charge transport.

The impact of variability sources on the 10 nm gate length NW-FET is addressed in two articles. Seoane et al. [2] investigated the impact of four major sources of intrinsic variability (line-edge roughness, gate-edge roughness, metal grain granularity in a gate, and random dopants in a transistor body) on the transistor performance in digital circuits. On the other hand, Li et al. [3] analysed the effect that metal gate work function fluctuations have on the transistor DC/AC characteristics with respect to different nanoscale metal grains and the variation of aspect ratio of channel cross-sections.

The effect of the impurities in the NW-FETs was also analysed in two articles. Sano et al. [4] focused their work on the physics associated with localized impurities inside the device, describing a systematic methodology on how to treat Coulomb interaction in many body-systems when using drift-diffusion simulations. Sady et al. [5], on the other hand, studied the effect of various scattering mechanisms and nanowire cross-section shapes on electron mobility in nanoscale Si NW-FETs.

In the article by Convertino et al. [6], the authors report on the fabrication of InGaAs based on FinFETs monolithically integrated on silicon substrates, presenting results for transistors with a gate length of 90 nm and a fin width of 40 nm. These InGaAs FinFETs could potentially replace the Si FinFET technology in low-power digital and RF applications.

Finally, Lee et al. [7], in their article, reviewed the theory regarding the lowest order approximation combined with Padé approaches for the quantum-mechanical treatment of electron–phonon and phonon–phonon inelastic scattering developed within the non-equilibrium Green's function (NEGF) formalism. The method was applied to the Si Gate-All-Around NW FET with a gate length of 13 nm. The NEGF formalism is very effective and thus a popular quantum transport technique to simulate carrier transport in very small quantum solid-state devices. The inclusion of inelastic scattering mechanisms into quantum transport techniques is very challenging, but essential to accurately account for the effect of self-heating and/or power dissipation in nanoscale semiconductor transistors because of their detrimental effect on the transistor performance and its reliability.

Conflicts of Interest: The authors declare no conflicts of interest.

References

1. Yoon, J.; Huang, F.; Shin, K.H.; Sohn, J.I.; Hong, W.-K. Effects of Applied Voltages on the Charge Transport Properties in a ZnO Nanowire Field Effect Transistor. *Materials* **2020**, *13*, 268. [CrossRef] [PubMed]
2. Seoane, N.; Nagy, D.; Indalecio, G.; Espiñeira, G.; Kalna, K.; García-Loureiro, A. A Multi-Method Simulation Toolbox to Study Performance and Variability of Nanowire FETs. *Materials* **2019**, *12*, 2391. [CrossRef] [PubMed]
3. Li, Y.; Chen, C.-Y.; Chuang, M.-H.; Chao, P.-J. Characteristic Fluctuations of Dynamic Power Delay Induced by Random Nanosized Titanium Nitride Grains and the Aspect Ratio Effect of Gate-All-Around Nanowire CMOS Devices and Circuits. *Materials* **2019**, *12*, 1492. [CrossRef] [PubMed]
4. Sano, N.; Yoshida, K.; Yao, C.-W.; Watanabe, H. Physics of Discrete Impurities under the Framework of Device Simulations for Nanostructure Devices. *Materials* **2018**, *11*, 2559. [CrossRef] [PubMed]
5. Sadi, T.; Medina-Bailon, C.; Nedjalkov, M.; Lee, J.; Badami, O.; Berrada, S.; Carrillo-Nunez, H.; Georgiev, V.; Selberherr, S.; Asenov, A. Simulation of the Impact of Ionized Impurity Scattering on the Total Mobility in Si Nanowire Transistors. *Materials* **2019**, *12*, 124. [CrossRef] [PubMed]
6. Convertino, C.; Zota, C.; Schmid, H.; Caimi, D.; Sousa, M.; Moselund, K.; Czornomaz, L. InGaAs FinFETs Directly Integrated on Silicon by Selective Growth in Oxide Cavities. *Materials* **2019**, *12*, 87. [CrossRef] [PubMed]
7. Lee, Y.; Logoteta, D.; Cavassilas, N.; Lannoo, M.; Luisier, M.; Bescond, M. Quantum Treatment of Inelastic Interactions for the Modeling of Nanowire Field-Effect Transistors. *Materials* **2020**, *13*, 60. [CrossRef] [PubMed]

Article

Effects of Applied Voltages on the Charge Transport Properties in a ZnO Nanowire Field Effect Transistor

Jongwon Yoon [1], Fu Huang [1], Ki Hoon Shin [2], Jung Inn Sohn [2,*] and Woong-Ki Hong [1,*]

[1] Jeonju Center, Korea Basic Science Institute, Jeonju-si, Jeollabuk-do 54907, Korea; jwyoon@kbsi.re.kr (J.Y.); hf3546@kbsi.re.kr (F.H.)

[2] Division of Physics and Semiconductor Science, Dongguk University-Seoul, Seoul 04620, Korea; kihoonshin@dongguk.edu

* Correspondence: junginn.sohn@dongguk.edu (J.I.S.); wkh27@kbsi.re.kr (W.-K.H.)

Received: 20 November 2019; Accepted: 31 December 2019; Published: 7 January 2020

Abstract: We investigate the effect of applied gate and drain voltages on the charge transport properties in a zinc oxide (ZnO) nanowire field effect transistor (FET) through temperature- and voltage-dependent measurements. Since the FET based on nanowires is one of the fundamental building blocks in potential nanoelectronic applications, it is important to understand the transport properties relevant to the variation in electrically applied parameters for devices based on nanowires with a large surface-to-volume ratio. In this work, the threshold voltage shift due to a drain-induced barrier-lowering (DIBL) effect was observed using a Y-function method. From temperature-dependent current-voltage (I-V) analyses of the fabricated ZnO nanowire FET, it is found that space charge-limited conduction (SCLC) mechanism is dominant at low temperatures and low voltages; in particular, variable-range hopping dominates the conduction in the temperature regime from 4 to 100 K, whereas in the high-temperature regime (150–300 K), the thermal activation transport is dominant, diminishing the SCLC effect. These results are discussed and explained in terms of the exponential distribution and applied voltage-induced variation in the charge trap states at the band edge.

Keywords: ZnO; nanowire; charge transport; field effect transistor; conduction mechanism

---

1. Introduction

Zinc oxide (ZnO) has received considerable interest over the past few decades as a promising material for a variety of applications in electronics, optics, and photonics because it exhibits a direct wide bandgap (~3.37 eV), a large exciton binding energy (60 meV), a variety of nanoscale forms, and piezoelectricity [1,2]. Recently, ZnO nanostructures have attracted much attention to the fields of nanoscale electronic and optoelectronic devices, such as sensors [3], solar cells [4], energy harvesting devices [5], light-emitting diodes [6], and especially field effect transistors (FETs) [7].

Since the FET based on nanowires is one of the fundamental building blocks in potential nanoelectronic applications, it is very important to understand charge transport behaviors in nanowire-based transistors. The electrical properties of nanowire-based FET devices sensitively depend on their size and shape, defects and impurities, and surface states or defects [7–9]. Moreover, it has been generally accepted that the contacts between the nanowire and the metal electrodes play also an important role in the charge transport properties of nanowire-based FETs due to their large surface-to-volume ratio coupled with unique geometry [10–12]. For example, Lee and coworkers reported the distinct electrical transport features of FETs made from ZnO nanowires with two different types of geometric properties: one type consisted of corrugated nanowires with a relatively smaller diameter and higher density of surface states or defects, and the other type involved smooth ZnO nanowires with a relatively larger diameter and lower density of surface states or defects [7]. Lord et al. [10] showed that the electrical transport behavior of nanocontacts between ZnO nanowires and Au metals can switch from Schottky to Ohmic depending on the size of the metal contact in relation to the nanowire diameter. Jo et al. [11] and He et al. [12] demonstrated the influence of the contact resistance on the electrical properties in In_2O_3 and ZnO nanowires, respectively.

In addition to structural geometry effects associated with nanowires and devices, importantly, a better understanding of the charge transport properties relevant to the variation in the electrical parameters actually applied to devices based on nanowires is required for the application of new nanoscale electronics and devices. Recently, several studies on the effect of bias stress in ZnO nanowire FETs have been reported [13,14]. Ju et al. [13] reported the effects of bias stress (gate or drain stress) on the stability of the ZnO nanowire FET with a self-assembled organic gate insulator. Choe et al. [14] investigated the threshold voltage instability induced by gate bias stress in ZnO nanowire FETs, which is associated with the trapping of charges in the interface trap sites located in interfaces between the nanowire and dielectric layer.

Herein, we report the effect of applied gate and drain voltages on the charge transport properties in a ZnO nanowire FET with a back-gated configuration. To do this, temperature-dependent current-voltage (I-V) measurements from 4 to 300 K were carried out. Using a *Y*-function method, we find that the threshold voltage (V_{th}) shifts to a negative gate bias direction due to the drain-induced barrier lowering (DIBL) effect, leading to increasing carrier concentration in the channel. The temperature-dependent I-V measurements show that the transport behavior of the fabricated ZnO nanowire FET is governed by space charge-limited conduction (SCLC) at low temperatures and low voltages, in particular by variable-range hopping (VRH) conduction mechanism in the temperature regime from 4 to 100 K, and by the thermal activation transport at the high-temperature regime (150–300 K).

2. Materials and Methods

High-density ZnO nanowires were grown on Au-coated c-plane sapphire substrates by a vapor transport method without using metal-catalysts. To grow the high-density ZnO nanowires, a mixed source of ZnO powder (99.995%) and graphite powder (99%) in a ratio of 1:1 was blended with ethanol. The source materials and substrates were placed in an alumina boat, which was then loaded into the center of a horizontal tube furnace. The furnace was heated at a rate of 35 °C/min and held at approximately 920 °C for 40–60 min. During the whole growth process, a mixed gas of Ar and O_2 with mixture ratio of 99:1 was maintained and then the flow rate of the mixed gas was 20 SCCM (standard cubic centimeters per minute) and the pressure of the furnace was kept at approximately 600 Torr. When the furnace was allowed to cool to room temperature naturally, a large amount of a white product was grown on the surface of the Au-coated c-plane sapphire substrate (not shown). Structural characterization of the ZnO nanowires vertically grown on the sapphire substrate was performed using field emission scanning electron microscope (FESEM) and transmission electron microscope (TEM), as shown in Figure S1. The energy dispersive x-ray spectroscopy (EDS) of the as-grown ZnO nanowires shows compositional elements (the inset in Figure S1a). The TEM images (Figure S1c–e) indicate that the growth direction of the ZnO nanowires is along the c-axis. A selected area electron diffraction (SAED) pattern confirms the (0001) growth direction (the inset of Figure S1d). The photoluminescence (PL) measurement of the ZnO nanowires at room temperature was examined by utilizing a FEX system (NOST, Seongnam-si, Korea) with a He–Cd laser (325 nm) as an incident excitation source (Figure S2). Next, the ZnO nanowires that were grown on the Au-coated sapphire substrate were transferred onto a highly-doped silicon wafer with 100 nm-thick thermally grown silicon dioxide (SiO_2) by dropping and drying a liquid suspension of ZnO nanowires for the fabrication of FET devices. For all the fabricated ZnO nanowire FETs, source and drain electrodes consisting of Ti (100 nm)/Au (80 nm) were deposited by an electron beam evaporator, as shown in Figure 1a. The distance between the source and drain electrodes is approximately 4 μm (Figure 1b). The electrical properties of the nanowire FET device were characterized using a semiconductor characterization system (Keithley 4200-SCS, Keithley, Cleveland, OH, USA) at a temperature range of 4–300 K. It should be noted that even though the nanowires are synthesized in the same conditions, there can be wire-to-wire or device-to-device variations in the electrical and optical properties, which strongly depend on the dimension (diameter and length, etc.) and surface states of the as-grown nanowires [7,15].

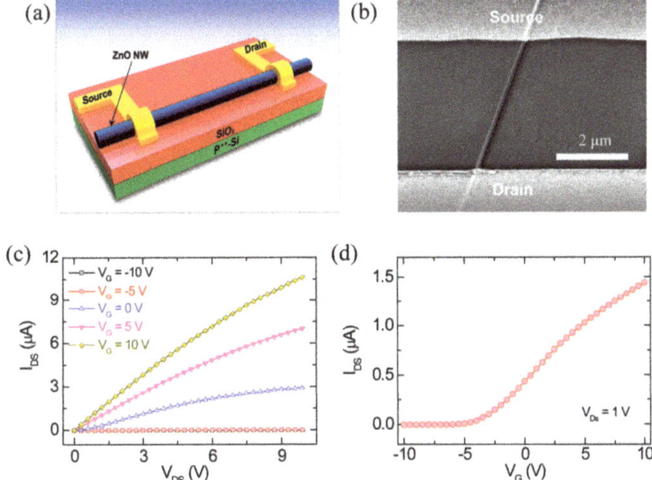

Figure 1. (**a**) Schematic illustration of the fabricated ZnO nanowire FET with a back-gate configuration; (**b**) A SEM image of the fabricated ZnO nanowire FET; (**c**) Output characteristics (I_{DS}-V_{DS}) and (**d**) transfer characteristics (I_{DS}-V_G) at V_{DS} = 1 V of the fabricated ZnO nanowire FET, which was measured at room temperature. The inset in (**d**) shows a semi-logscale I_{DS}-V_G curve at V_{DS} = 1 V.

3. Results and Discussion

A schematic illustration and a scanning electron microscopy (SEM) image of the fabricated ZnO nanowire FET with a back-gate configuration are shown in Figure 1a,b. Figure 1c,d shows the output (I_{DS}-V_{DS}) and transfer (I_{DS}-V_G) characteristics of the fabricated ZnO nanowire FET with a back-gate configuration (Figure 1a,b), respectively. The fabricated ZnO nanowire FET showed typical n-type semiconductor properties and depletion-mode operation, which exhibited a nonzero current at zero gate bias and a negative threshold voltage [15].

Figure 2a shows the transfer characteristics at different drain-source voltages for the fabricated ZnO nanowire FET measured at room temperature. From this, electrical characteristics were analyzed by the *Y*-function method (YFM) (Figure 2b), which has been widely used for contact resistance and mobility based on a straightforward analysis of the drain current (I_{DS}) in the linear region (electron accumulation region) [16,17]. The Y-function can be obtained from the I_{DS}-V_G (Figure 2a) as follows [17],

$$Y = \frac{I_{DS}}{\sqrt{g_m}} = \sqrt{V_{DS}\frac{\mu C_G}{L^2}(V_G - V_{th})} \tag{1}$$

where $g_m = dI_{DS}/dV_G$, μ is the mobility, C_G is the gate capacitance, L is the channel length, and V_{th} is the threshold voltage, in which μ and V_{th} can be determined from the slope and the V_G-axis intercept of the linear region of the *Y*-function, respectively (Figure 2b,c). In Figure 2b, it is clearly seen that V_{th} shifts to a negative gate bias direction (marked by arrows) when V_{DS} increases from 0.5 to 2.5 V, which indicates the DIBL effect [18]. This effect can reduce the Schottky barrier between source/drain electrodes and the nanowire contacts, affecting the contact resistance (R_C). Using the *Y*-function, the R_C at interfaces between source/drain electrodes and the ZnO nanowire can be calculated from the following equation [17],

$$R_C = R_{tot} - R_{ch} = \frac{V_{DS}}{I_{DS}} - \frac{V_{DS}}{k^2(V_{GS} - V_{th})} \tag{2}$$

where k is the slope of the linear region of the Y-function. The slopes of the linear region of the Y-function are different (Figure 2c), indicating the difference in R_C [17] (Figure 2d). Importantly, the contact resistance is present at a metal-nanowire interface and can affect the electrical performance of nanowire FETs [19]. The work function difference between the ZnO and the contact metal leads to the formation of an energy barrier at the interface between the two materials, which can influence the barrier height.

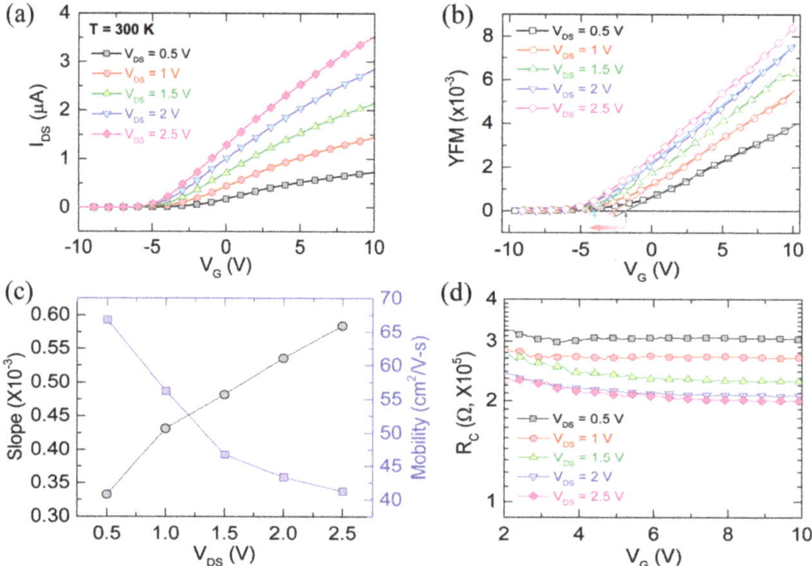

Figure 2. (**a**) I_{DS}-V_G curves measured at room temperature (T = 300 K) for the ZnO nanowire FET, with V_{DS} varying from 0.5 to 2.5 V; (**b**) YFM value as a function of V_G at different V_{DS} values for the ZnO nanowire FET. From the linear fitting, V_{th} and mobility can be extracted from the V_G-axis intercept and the slope, respectively. Each arrow indicates the V_{th} for each V_{DS}; (**c**) Slope and mobility as a function of V_{DS} extracted from linearly fitted curves in (**b**); (**d**) Contact resistance as a function of gate bias, with V_{DS} varying from 0.5 to 2.5 V.

To understand the charge transport mechanism in our nanowire FET with different contact resistances, the temperature-dependent electrical measurement and analyses of the ZnO nanowire FET were examined. Figure 3a shows the I_{DS}-V_{DS} characteristics of the ZnO nanowire FET at different temperatures ranging from 30 to 200 K. With decreasing temperature, the I_{DS} decreased, indicating a strong temperature dependence. In addition, the logscale I_{DS}–V_{DS} showed the power law relationship, $I \propto V^{\alpha}$, and such power law dependence with $\alpha > 2$ is a characteristic feature of SCLC in a semiconductor with an exponential charge trap distribution at the band edge [19,20]. The exponents, α, were extracted from logscale I_{DS}-V_{DS} curves in the temperature range from 4 to 300 K at different gate biases, as shown in the inset of Figure 3b. The α values increased with decreasing temperature, exceeding 2 in the low-temperature range. This result implies the existence of trap states in the ZnO nanowire. The values reached approximately 1 in the high- temperature range due to the thermally activated electrons, resulting in deviation from SCLC. The trap densities (N_t) can be estimated by extrapolating the logscale I_{DS}-V_{DS} characteristics, as shown in Figure 3b. Figure 3b shows a crossover point at which the conductance was independent of the temperature. The V_{DS} value at the crossover point is denoted as a crossover voltage (V_c) and it was approximately 25.4 V. The V_c can be expressed by [20],

$$V_C = \frac{qN_tL^2}{2\varepsilon_0\varepsilon_r} \tag{3}$$

where q is the electric charge, L is the channel length, ε_0 is the vacuum permittivity, and ε_r is the relative permittivity of ZnO (~8.5). From the above equation, the calculated N_t at V_c = 25.4 V, was 1.5×10^{15} cm^{-3}. According to previous reports [21,22], most of the trap densities arise from oxygen vacancies located on the nanowire surface rather than at the nanowire center. Therefore, the calculated N_t may correspond to the interface trap states at the metal-nanowire contacts or the nanowire-dielectric layer, which could affect the charge transport of the ZnO nanowire FET.

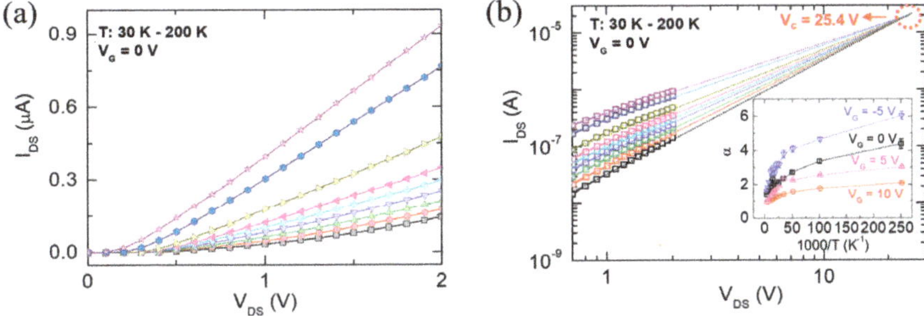

Figure 3. (**a**) I_{DS}-V_{DS} curves measured at V_G = 0 V and different temperatures (30–200 K) for the ZnO nanowire FET; (**b**) The extrapolation derived from the corresponding logscale I_{DS}-V_{DS} characteristics at different temperatures of (**a**), which provide a critical voltage (V_C).

Next, we carried out analyses of the Arrhenius plots of the conductance (G) versus 1000/T at different V_G values to further investigate the transport mechanism of the ZnO nanowire FET, as shown in Figure 4a,b. Two different regimes in the temperature-dependent conductance of the nanowire FET device were clearly observed at different V_G values (V_G from −3 to 10 V, 1 V steps), implying different charge transport mechanisms. Note that the Arrhenius plots at low V_{DS} regime (0.5, 1, 1.5, and 2 V) were also characterized for different V_G values. In the high-temperature region (150–300 K) (marked by the gray-colored region), the thermally activated carriers were dominant in the charge transport, indicating a conductance proportional to $exp(-E_a/kT)$, which can be expressed as Equation (4) below [23–25].

$$G = G_0 exp\left(-\frac{E_a}{k_BT}\right) \tag{4}$$

where G and G_0 are the conductance and weak temperature-dependent constant, respectively, E_a is the activation energy, k_B is the Boltzmann constant, and T is the temperature. The E_a characterized by the linear region in the semi-log plot of conductance versus $1/T$ is shown in Figure 4a. Here, the E_a can be extracted by the linear fits in the high-temperature region in Figure 4a (marked by the gray-colored region). Figure 4c shows the extracted E_a as a function of the V_G at different V_{DS} values for the device. The E_a decreased due to the lowered Schottky barrier at the metal/semiconductor interface when the applied biases increased, including V_G and V_{DS}. In contrast, in the low-temperature region (4–100 K), the carrier conduction is mainly attributed to VRH, which exhibits charge transport through the trap states near the Fermi level. According to previous reports [23,26–30], the VRH conduction can be expected due to charge trapping at localized states in semiconducting nanomaterials at low applied bias and low temperature where the Fermi level lies in localized sates within a band gap.

The conductance following the three-dimensional (3D) VRH mechanism can be expressed by the following equation [25,31,32],

$$G = G_0 exp\left[-\left(\frac{T_0}{T}\right)^{1/4}\right] \tag{5}$$

where T_0 are the characteristic characteristic temperature. Figure 4b shows that the low-temperature conductance of the device is well fitted by the 3D VRH as a function of $T^{-1/4}$ at low applied bias, indicating that the conductance follows 3D VRH model well for low electric fields. From Equation (5), the values of T_0, which represent how actively VRH occurs [25,31,32], were extracted, as shown in Figure 4d. As the applied biases (V_G and low V_{DS}) increased, the T_0 also continuously decreased, implying reduced VRH conduction. The result might be due to the enhanced electron concentration from the lowering of the Schottky barrier. The increased electron concentration might additionally fill the trap states, leading to the reduction in hopping conduction [25,31,32]. As a result, the E_a and T_0 values can be modified by the applied electric field, which is associated with the modulation of localized trap states. This trend is consistent with the results reported for semiconducting nanomaterials with localized trap states [23,30,33].

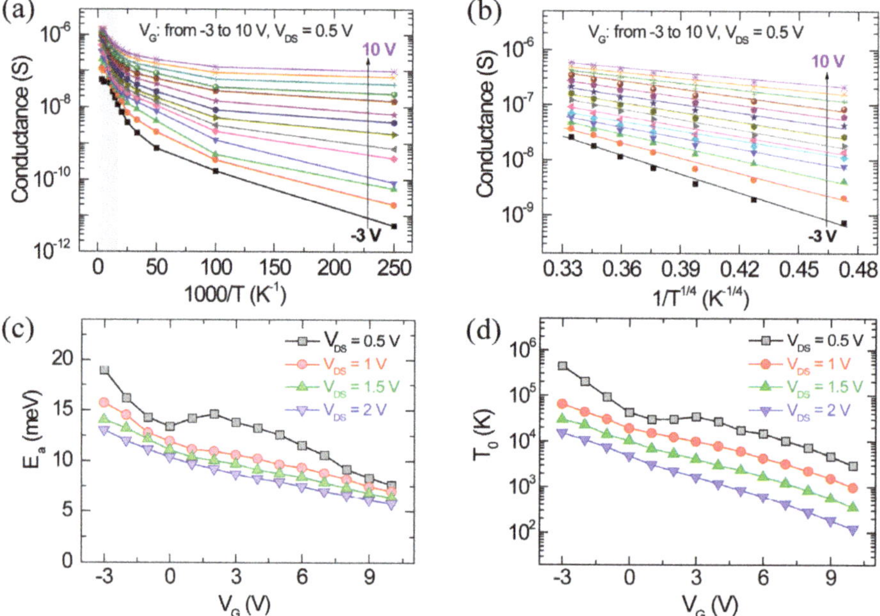

Figure 4. (**a**) Arrhenius plots of the conductance (*G*) versus *1000/T* at different gate voltages from −3 to 10 V for V_{DS} = 0.5 V. (**b**) Semilogarithm plots showing the temperature dependence of conductance (*G*) vs $1/T^{1/4}$ fitted by Equation (5) at different gate voltages for V_{DS} = 0.5 V. The activation energy (E_a) (**c**) and characteristic temperature (T_0) (**d**) depending on the applied gate and drain voltages.

The energy band diagram presented in Figure 5 qualitatively shows the charge transport mechanisms of the ZnO nanowire FET, as discussed above. Unlike the equilibrium condition (Figure 5a), the applied biases (V_G and V_{DS}) could induce Schottky barrier modulation, resulting in changes in the carrier injection properties at the metal-semiconductor contact, as shown in Figure 5b. As a result, the modified Schottky barrier could affect the carrier concentration, leading to a change in the density of localized trap states in the channel. Furthermore, different temperature-dependent charge transport mechanisms were observed. Specifically, thermal activated (TA) conduction of electrons from a shallow level of localized states was dominant for charge transport in the high-temperature range, denoted as TA in Figure 5b (left), whereas the VRH conduction through the trap states near the Fermi level was dominant in the low-temperature range, denoted as VRH in Figure 5b (right).

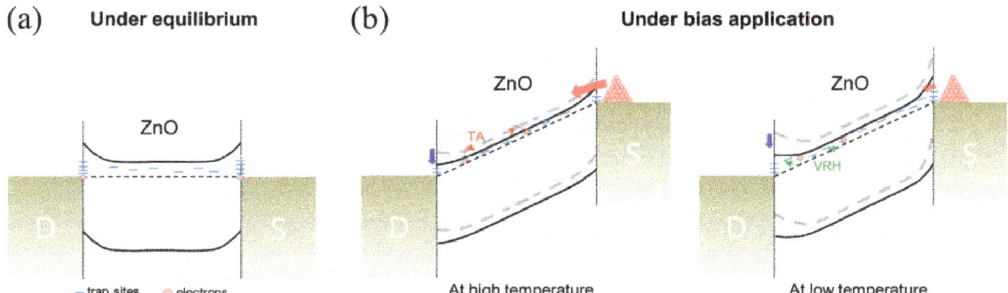

Figure 5. Energy band diagrams depicting the charge transport mechanism for the ZnO nanowire FET (**a**) under equilibrium and (**b**) under bias application at low and high temperatures. The blue arrow indicates Schottky barrier modulation according to the applied gate and drain voltages.

4. Conclusions

In summary, we fabricated a ZnO nanowire FET with a back-gated configuration and characterized the electrical properties of the FET device through temperature-dependent measurements to study the effect of applied gate and drain voltages on the charge transport properties. The Y-function method showed that the V_{th} shifted to a negative gate bias direction due to the DIBL effect. The temperature-dependent I-V measurements showed that the transport behavior of the ZnO nanowire FET was governed by SCLC at low temperatures and low voltages, in particular, by VRH conduction in the temperature regime from 4 to 100 K and by thermal activation transport at the high-temperature regime (150–300 K).

Supplementary Materials: The following are available online at http://www.mdpi.com/1996-1944/13/2/268/s1, Figure S1: SEM and TEM characterizations of the as-grown ZnO nanowires, Figure S2: PL data of the as-grown ZnO nanowires.

Author Contributions: Experiments; J.Y., F.H. and K.H.S.; writing—original draft preparation, review and editing; J.Y., J.I.S. and W.-K.H.; supervision, J.I.S. and W.-K.H. All authors have read and agreed to the published version of the manuscript.

Funding: This research was funded by Korea Basic Science Institute (grant numbers D39630 and D010500). J.I.S. acknowledges that this work was supported by the Dongguk University Research Fund of 2018.

Conflicts of Interest: The authors declare no conflict of interest.

References

1. Wang, Z.L. Zinc oxide nanostructures: Growth, properties and applications. *J. Phys. Condens. Matter* **2004**, *16*, R829–R858. [CrossRef]
2. Yang, P.; Yan, H.; Mao, S.; Russo, R.; Johnson, J.; Saykally, R.; Morris, N.; Pham, J.; He, R.; Choi, H.-J. Controlled growth of ZnO nanowires and their optical properties. *Adv. Funct. Mater.* **2002**, *12*, 323–331. [CrossRef]
3. Rackauskas, S.; Barbero, N.; Barolo, C.; Viscardi, G. ZnO nanowire application in chemoresistive sensing: A review. *Nanomaterials* **2017**, *7*, 381. [CrossRef] [PubMed]
4. Consonni, V.; Briscoe, J.; Kärber, E.; Li, X.; Cossuet, T. ZnO nanowires for solar cells: A comprehensive review. *Nanotechnology* **2019**, *30*, 362001. [CrossRef] [PubMed]
5. Li, Y.; Çelik-Buter, Z.; Butler, D.P. An integrated piezoelectric zinc oxide nanowire micro-energy harvester. *Nano Energy* **2016**, *26*, 456–465. [CrossRef]
6. Bao, J.; Zimmler, M.A.; Capasso, F.; Wang, X.; Ren, Z.F. Broadband ZnO single-nanowire light-emitting diode. *Nano Lett.* **2006**, *6*, 1719–1722. [CrossRef]

7. Hong, W.-K.; Sohn, J.I.; Hwang, D.-K.; Kwon, S.-S.; Jo, G.; Song, S.; Kim, S.-M.; Ko, H.-J.; Park, S.-J.; Welland, M.E.; et al. Tunable electronic Transport Characteristics of Surface-Architecture-Controlled ZnO Nanowire Field Effect Transistors. *Nano Lett.* **2008**, *8*, 950–956. [CrossRef] [PubMed]
8. Lieber, C.M. One- dimensional nanostructures: Chemistry, physics & applications. *Solid State Commun.* **1998**, *107*, 607–616.
9. Schmidt-Mende, L.; MacManus-Driscoll, J.L. ZnO—Nanostructures, defects, and devices. *Mater. Today* **2007**, *10*, 40–48. [CrossRef]
10. Lord, A.M.; Maffeis, T.G.; Kryvchenkova, O.; Cobley, R.J.; Kalna, K.; Kepaptsoglou, D.M.; Ramasse, Q.M.; Walton, A.S.; Ward, M.B.; Köble, J.; et al. Controlling the electrical transport properties of nanocontacts to nanowire. *Nano Lett.* **2015**, *15*, 4248–4254. [CrossRef]
11. Jo, G.; Maeng, J.; Kim, T.-W.; Hong, W.-K.; Choi, B.-S.; Lee, T. Channel-length and gate-bias dependence of contact resistance and mobility for In_2O_3 nanowire field effect transistors. *J. Appl. Phys.* **2007**, *102*, 084508. [CrossRef]
12. He, J.-H.; Ke, J.-J.; Chang, P.-H.; Tsai, K.-T.; Yang, P.C.; Chan, I.-M. Development of Ohmic nanocontacts via surface modification for nanowire-based electronic and optoelectronic devices: ZnO nanowires as an example. *Nanoscale* **2012**, *4*, 3399–3404. [CrossRef] [PubMed]
13. Ju, S.; Janes, D.B.; Lu, G.; Facchetti, A.; Marks, T.J. Effects of bias stress on ZnO nanowire field-effect transistors fabricated with organic gate nanodielectrics. *Appl. Phys. Lett.* **2006**, *89*, 193506. [CrossRef]
14. Choe, M.; Park, W.; Kang, J.W.; Jeong, S.; Hong, W.-K.; Lee, B.H.; Park, S.-J.; Lee, T. Investigation of threshold voltage instability induced by gate bias stress in ZnO nanowire field effect transistors. *Nanotechnology* **2012**, *23*, 485201. [CrossRef]
15. Hong, W.-K.; Hwang, D.-K.; Park, I.-K.; Jo, G.; Song, S.; Park, S.-J.; Lee, T.; Kim, B.-J.; Stach, E.A. Realization of highly reproducible ZnO nanowire field effect transistors with n-channel depletion and enhancement modes. *Appl. Phys. Lett.* **2007**, *90*, 243103. [CrossRef]
16. Kim, T.-Y.; Amani, M.; Ahn, G.H.; Song, Y.; Javey, A.; Chung, S.; Lee, T. Electrical properties of synthesized large-area MoS_2 field-effect transistors fabricated with inkjet-printed contacts. *ACS Nano* **2016**, *10*, 2819–2826. [CrossRef]
17. Zhao, Y.; Xiao, X.; Huo, Y.; Wang, Y.; Zhang, T.; Jiang, K.; Wang, J.; Fan, S.; Li, Q. Influence of asymmetric contact form on contact resistance and Schottky barrier, and corresponding applications of diode. *ACS Appl. Mater. Interfaces* **2017**, *9*, 18945–18955. [CrossRef]
18. Jo, G.; Maeng, J.; Kim, T.-W.; Hong, W.-K.; Jo, M.; Hwang, H.; Lee, T. Effects of channel-length scaling on In_2O_3 nanowire field effect transistors studied by conducting atomic force microscopy. *Appl. Phys. Lett.* **2007**, *90*, 173106. [CrossRef]
19. Liao, Z.-M.; Lv, Z.-K.; Zhou, Y.-B.; Xu, J.; Zhang, J.-M.; Yu, D.-P. The effect of adsorbates on the space–charge-limited current in single ZnO nanowires. *Nanotechnology* **2008**, *19*, 335204. [CrossRef]
20. Xu, W.; Chin, A.; Ye, L.; Ning, C.Z.; Yu, H. Charge transport and trap characterization in individual GaSb nanowires. *J. Appl. Phys.* **2012**, *111*, 104515. [CrossRef]
21. Fang, D.Q.; Zhang, R.Q. Size effects on formation energies and electronic structures of oxygen and zinc vacancies in ZnO nanowires: A first-principle study. *J. Appl. Phys.* **2011**, *109*, 044306. [CrossRef]
22. Shao, Y.; Yoon, J.; Kim, H.; Lee, T.; Lu, W. Analysis of surface states in ZnO nanowire field effect transistors. *Appl. Surf. Sci.* **2014**, *301*, 2–8. [CrossRef]
23. Chang, P.-C.; Lu, J.G. Temperature dependent conduction and UV induced metal-to-insulator transition in ZnO nanowires. *Appl. Phys. Lett.* **2008**, *92*, 212113. [CrossRef]
24. Lin, Y.-F.; Jian, W.-B.; Wang, C.P.; Suen, Y.-W.; Wu, Z.-Y.; Chen, F.-R.; Kai, J.-J.; Lin, J.-J. Contact to ZnO and intrinsic resistances of individual ZnO nanowires with a circular cross section. *Appl. Phys. Lett.* **2007**, *90*, 223117. [CrossRef]
25. Lin, Y.-F.; Chen, T.-H.; Chang, C.-H.; Chang, Y.-W.; Chiu, Y.-C.; Hung, H.-C.; Kai, J.-J.; Liu, Z.; Fang, J.; Jian, W.-B. Electron transport in high-resistance semiconductor nanowires through two-probe measurements. *Phys. Chem. Chem. Phys.* **2010**, *12*, 10928–10932. [CrossRef] [PubMed]
26. Ko, D. Charge Transport Properties in Semiconductor Nanowires. Ph.D. Thesis, Ohio State University, Columbus, OH, USA, 2011.

27. Liu, X.; Liu, J.; Antipina, L.Y.; Hu, J.; Yue, C.; Sanchez, A.M.; Sorokin, P.B.; Mao, Z.; Wei, J. Direct fabrication of functional ultrathin single-crystal nanowires from quasi-one-dimensional van der Waals crystals. *Nano Lett.* **2016**, *16*, 6188–6195. [CrossRef]
28. Kaiser, A.B.; Gómez-Navarro, C.; Sundaram, R.S.; Burghard, M.; Kern, K. Electrical conduction mechanism in chemically derived graphene monolayers. *Nano Lett.* **2009**, *9*, 1787–1792. [CrossRef]
29. Varade, V.; Anjaneyulu, P.; Sangeeth, C.S.S.; Ramesh, K.P.; Menon, R. Electric field activated nonlinear anisotropic charge transport in doped polypyrrole. *Appl. Phys. Lett.* **2013**, *103*, 233305. [CrossRef]
30. Radisavljevic, B.; Kis, A. Mobility engineering and a metal-insulator transition in monolayer MoS_2. *Nat. Mater.* **2013**, *12*, 815–820. [CrossRef] [PubMed]
31. Tian, J.; Cai, J.; Hui, C.; Zhang, C.; Bao, L.; Gao, M.; Shen, C.; Gao, H. Boron nanowires for flexible electronics. *Appl. Phys. Lett.* **2008**, *93*, 122105. [CrossRef]
32. Ma, Y.-J.; Zhang, Z.; Zhou, F.; Lu, L.; Jin, A.; Gu, C. Hopping conduction in single ZnO nanowires. *Nanotechnology* **2005**, *16*, 746–749. [CrossRef]
33. Ayari, A.; Cobas, E.; Ogundadegbe, O.; Fuhrer, M.S. Realization and electrical characterization of ultrathin crystals of layered transition-metal dichalcogenides. *J. Appl. Phys.* **2007**, *101*, 014507. [CrossRef]

Article

A Multi-Method Simulation Toolbox to Study Performance and Variability of Nanowire FETs

Natalia Seoane [1,*], Daniel Nagy [1], Guillermo Indalecio [1], Gabriel Espiñeira [1], Karol Kalna [2] and Antonio García-Loureiro [1]

[1] Centro Singular de Investigación en Tecnoloxías da Información, University of Santiago de Compostela, 15782 Santiago de Compostela, Spain

[2] Nanoelectronic Devices Computational Group, College of Engineering, Swansea University, Swansea, Wales SA1 8EN, UK

* Correspondence: natalia.seoane@usc.es; Tel.: +34-881-816-424

Received: 10 June 2019; Accepted: 19 July 2019; Published: 26 July 2019

Abstract: An in-house-built three-dimensional multi-method semi-classical/classical toolbox has been developed to characterise the performance, scalability, and variability of state-of-the-art semiconductor devices. To demonstrate capabilities of the toolbox, a 10 nm gate length Si gate-all-around field-effect transistor is selected as a benchmark device. The device exhibits an off-current (I_{OFF}) of 0.03 μA/μm, and an on-current (I_{ON}) of 1770 μA/μm, with the I_{ON}/I_{OFF} ratio 6.63×10^4, a value 27% larger than that of a 10.7 nm gate length Si FinFET. The device SS is 71 mV/dec, no far from the ideal limit of 60 mV/dec. The threshold voltage standard deviation due to statistical combination of four sources of variability (line- and gate-edge roughness, metal grain granularity, and random dopants) is 55.5 mV, a value noticeably larger than that of the equivalent FinFET (30 mV). Finally, using a fluctuation sensitivity map, we establish which regions of the device are the most sensitive to the line-edge roughness and the metal grain granularity variability effects. The on-current of the device is strongly affected by any line-edge roughness taking place near the source-gate junction or by metal grains localised between the middle of the gate and the proximity of the gate-source junction.

Keywords: nanowire field-effect transistors; variability effects; Monte Carlo; Schrödinger based quantum corrections; drift-diffusion

---

1. Introduction

Gate-all-around nanowire field-effect transistors (GAA-NW FETs) are one of the main contenders for future CMOS technologies [1] since they provide a better electrostatic control of the channel when compared to fin field-effect transistors (FinFETs) [2], the current architecture adopted by the semiconductor industry. In addition, nanowire based transistor architectures extend beneficial properties of multi-gate devices required in digital circuits such as quasi-1D current transport, largely confined electrical fields, and immunity of threshold voltage from substrate bias [3].

On the other hand, GAA-NW FETs, like all deeply scaled semiconductor devices, are greatly affected by variability issues [4], related to either the fabrication process or material variations that can limit their performance and reliability [5]. Previous studies have shown that in the sub-threshold region, GAA-NW FETs are less resilient to intrinsic sources of variability than FINFETs [6,7]. In addition, the significant degradation observed in the on-region performance of GAA-NW FETs due to line edge roughness variations could be a critical issue for the scaling of these devices [6].

Nowadays, technology computer-aided design (TCAD) tools play a key role in the advancement of the semiconductor industry [8]. The TCAD tools are able to quickly characterise semiconductor devices, not only fabricated but also foreseen, and allow to investigate the impact of changes in materials, designs or fabrication processes. Currently, three-dimensional (3-D) simulations are necessary to appropriately model devices such as FinFETs or GAA-NW, due to the two-dimensional (2-D) nature of the quantum confinement, which increases the computational cost of a study [9]. There are different approaches that can be used in simulations of state-of-the-art semiconductor devices, ranging from the relatively simple and low computationally demanding drift-diffusion method [10], to extremely complex quantum mechanical approaches, such as pseudopotential-based electron quantum transport [11] or the non-equilibrium Green's functions (NEGF) [12] formalism, that can also be coupled to empirical tight-binding models [13]. The use of fully quantum simulators is computationally prohibitive for statistical studies, being essential a trade-off between the simulation's accuracy and the calculation time.

In this work, an in-house built finite-element multi-method semi-classical/classical simulation toolbox acronymed VENDES (Variability Enabled Nanometric DEvice Simulator) is used to characterise nano-scaled semiconductor devices including their operational performance and variability. To demonstrate capabilities of the VENDES, the performance and variability of a 10 nm gate length Si GAA-NW FET scaled down from an experimental device [14] is studied and assessed. The paper is organized as follows: the simulation techniques available in VENDES used in this study are described in Section 2. Section 3 analyses the performance and resilience to variability of the 10 nm gate length Si GAA-NW FET. Finally, Section 4 draws the main conclusions of this work.

2. Simulation Framework

VENDES, a 3D finite-element (FE) based device simulator, has been developed jointly at Universidade de Santiago de Compostela (Spain) and at Swansea University (United Kingdom). Figure 1 shows the basic flowchart of the VENDES toolbox.

The starting point is the generation of the FE mesh via the open source software Gmsh [15]. The FE method allows not only an accurate description of complex simulation domains, as in the case of elliptic cross-section shaped GAA-NW FETs [16], but also the possibility of introducing realistic deformations to the device dimensions. This capability of accurate geometrical description is crucial in the modelling of variability effects because a correct distribution of potential and carrier density is essential to predict the experimentally observed behaviour. Note that, for devices deeply scaled into the nanometre regime, the size of device variations and deformations can be comparable to critical device dimensions. The sources of variability included in VENDES either alter the structure dimensions or modify some physical properties affecting the device nodes and are described in detail in Section 3.2.

The classical electrostatic potential, V_{cl}, is obtained from the Poisson equation solution on every node of the 3D FE tetrahedral mesh:

$$div(\varepsilon(r)\nabla V_{cl}(r)) = q(p(r) - n(r) + N_D^+(r) - N_A^-(r)), \quad (1)$$

where $r = (x, y, z)$ is the spatial coordinate, $\varepsilon(r)$ is the dielectric constant of the material, $n(r)$ and $p(r)$ are the electron and hole densities and $N_D^+(r)$ and $N_A^-(r)$ are the effective doping concentrations of donors and acceptors, respectively.

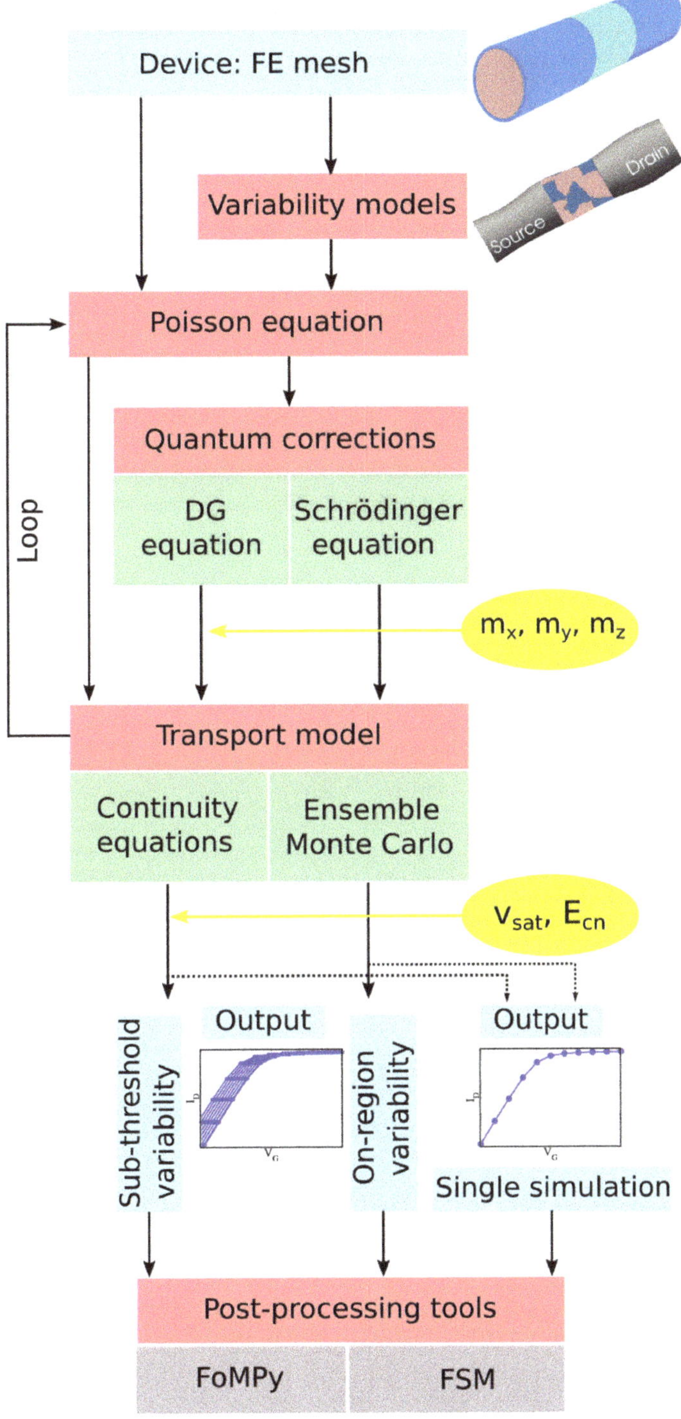

Figure 1. Basic flowchart of VENDES (Variability Enabled Nanometric DEvice Simulator).

Quantum corrections are incorporated in VENDES via two different techniques: i) the 3D density gradient (DG) equation and ii) the 2D Schrödinger (SCH) equation. In the first case, the DG quantum potential for electrons, $V_{dg}(r)$ [17], is obtained as:

$$V_{dg}(r) = 2[b_n]\frac{\nabla^2\sqrt{n(r)}}{\sqrt{n(r)}} = \phi_n(r) - V_{cl}(r) + \frac{k_B T}{q}\ln\left(\frac{n(r)}{n_i(r)}\right), \tag{2}$$

where

$$[b_n] = \frac{\hbar^2}{4qr_n}\begin{pmatrix} 1/m_x & 0 & 0 \\ 0 & 1/m_y & 0 \\ 0 & 0 & 1/m_z \end{pmatrix}. \tag{3}$$

Here, $\phi_n(r)$ is the quasi-Fermi potential for electrons, $n_i(r)$ is the intrinsic carrier concentration of electron and holes, k_B is the Boltzmann constant, T is the lattice temperature, $\hbar$ is the reduced Planck constant, r_n is a dimensionless parameter that models statistical phenomena [18] and m_x, m_y and m_z are the DG electron effective masses in the x-, y- and z-directions, respectively [17]. It is important to remark that these effective masses operate as fitting parameters [19] and are not related to the material transport effective masses. The DG effective masses in the transverse directions (m_y and m_z) will account for the strength of the quantum-mechanical confinement of the carriers in the device channel through a threshold voltage shift [20]. The DG effective mass in the transport direction (m_x) can account for the source-to-drain tunnelling by lowering the barrier of classical electrostatic potential which occurs between the source and the drain when the transistor is operating in the sub-threshold region [20]. The main drawback of the DG based quantum corrections is that they require calibration against either experimental data (when available) or more complex simulation techniques (such as Monte Carlo or Non-Equilibrium Green's Functions) [21].

On the other hand, the quantum correction method based on the solution of the Schrödinger equation [22] is calibration free. This technique assumes longitudinal and transverse electron effective masses in a minimum of the conduction valley of silicon and accounts for wave-functions penetrating into a surrounding dielectric layer [16,23]. The SCH equation is solved on two-dimensional (2D) slices placed across the device channel using a non-uniform distribution dependent on the gradient of electron density. The 2D quantum-mechanical electron density, $n_{sc}(y,z)$, is obtained from the SCH equation eigen-states, $\psi_i(y,z;E_i)$, and their corresponding eigen-energies, E_i, as follows:

$$n_{sc}(y,z) = g\frac{\sqrt{2\pi m^* k_B T}}{\pi\hbar}\sum_i |\psi_i(y,z;E_i)|^2 \exp\left[\frac{E_{F_n} - E_i}{k_B T}\right], \tag{4}$$

where E_{F_n} is the electron quasi-Fermi level, g the degeneracy factor, and m^* the electron effective transport mass. The Equation (4) considers Boltzmann statistics and assumes six equivalent valleys for Si ($g = 6$). The SCH quantum correction can be considered 'isotropic', when the electron effective mass in silicon is taken to be average of longitudinal and transverse electron effective masses, but also 'anisotropic', when longitudinal and transverse electron effective masses that are dependent on the valley orientation are considered. In that case, Equation (4) is solved separately for each of the three Δ valleys, taking into account the different sub-band edges (i.e., appropriate energy levels) for the different valleys, obtaining a different $n_{sc}(y,z)$ for each valley ($g = 2$) and m^* will be dependent on the channel orientation which can be $\langle 100\rangle$ or $\langle 110\rangle$ as shown in [24].

The electron density, $n_{sc}(y,z)$, calculated on the 2D slices, is interpolated to a 3D device density domain to obtain $n_{sc}(r)$. The resulting SCH quantum correction potential, $V_{sc}(r)$ [22], is as follows:

$$V_{sc}(r) = \frac{k_B T}{q}\log(n_{sc}(r)/n_i(r)) - V_{cl}(r). \tag{5}$$

Note that, in the anisotropic SCH quantum correction, a separate $V_{sc}(r)$ is obtained for each valley.

To simulate the transport inside the channel of the device, VENDES has implemented two different carrier transport methods: i) the drift-diffusion (DD) approach and ii) an ensemble Monte Carlo (MC) technique. The DD approach couples the electrostatic potential obtained from the quantum corrected solution of Poisson equation with the current continuity equation for electrons in order to obtain the electron current density, $J_n(r)$, as:

$$J_n(r) = -q\mu_n(r)n(r)\nabla(\phi_n(r)), \tag{6}$$

$$div(J_n(r)) = qR(r), \tag{7}$$

where $\mu_n(r)$ is the electron mobility and $R(r)$ is the recombination term (set to zero by default). Note that, the DD method only accounts for the local relationship between the velocity and the electric field and it is unable to correctly represent non-equilibrium transport effects [25]. However, some of the non-equilibrium phenomena can be partially mimicked via appropriate mobility models. To model the carrier transport behaviour in GAA-NW FETs, VENDES uses Caughey-Thomas doping dependent low-field electron mobility model [26] coupled with perpendicular and lateral electric field models [27] which better describe carrier transport at large electric fields. When using these mobility models, the main calibration parameters are a low-field carrier mobility, a critical electric field, E_{cn}, and a saturation velocity, v_{sat}.

The limitations of the DD approach can be overcome by using a semi-classical transport model, the MC technique where an ensemble of particles representing carriers evolves through free flights governed by Newton equations and undergoes scattering events with a probability which is determined quantum-mechanically. The best way to initialise the distribution of carriers in the real space, is to use the quantum corrected potential from the solution of the Poisson's equation, which results in speeding up the simulation. MC uses the analytical non-parabolic anisotropic approximation [28] for the silicon band structure taking into account three valley minima, X, L, and Γ, using Herring-Vogt transformation to transform ellipsoidal surfaces to spherical ones in order to simplify a calculation of free flights and scattering events. The MC technique considers carrier scattering in a quantum-mechanical way by using typically Fermi Golden Rule [29] to obtain the transition rates. The following electron scattering mechanisms, important for silicon based devices, are included in VENDES: i) electron interactions with intra- and inter-valley acoustic and non-polar optical phonons [28,30], ii) electron interactions with ionised impurities using Ridley's third body exclusion [31,32] and static screening [29], and iii) electron interaction with interface roughness using Ando's 2D potential approach [33]. VENDES uses Boltzmann statistics when solving 3D Poisson equation and determining a final state after electron scattering but, the electron scattering with ionised impurities uses Fermi-Dirac statistics to calculate the static screening by a self-consistent calculation of the Fermi energy and the electron temperature from the average electron density and kinetic energy in a whole real space device simulation domain at each scattering event [34]. The inclusion of Fermi-Dirac statistics into electron scattering with ionised impurities turns to be sufficient to correctly simulate injection of carriers into the channel from a heavily doped source/drain when comparing the results from quantum corrected 3D finite element Monte Carlo device simulations with experimentally measured I-V characteristics in nanoscale FinFETs [29] and nanowire FETs [16].

3. Performance and Variability of GAA-NW FETs

In this work, VENDES has been applied to study state-of-the-art nanoscale GAA-NW FETs designed for future digital technology node generations [35]. Section 3.1 presents the GAA-NW FET description and main figures of merit. Section 3.2 shows a thorough analysis of the impact that different sources of fluctuations have on this architecture.

3.1. Benchmark Device

The device under study is a 10 nm gate length Si GAA-NW FET with an elliptically shaped cross-section that has been scaled [16] following the ITRS guidelines [35] from an experimental 22 nm gate length device from IBM [14]. Ref. [14] includes TEM images of the fabricated structure and the I_D-V_G characteristics that we have used to validate our device. The elliptically shaped cross-section of the transistor body as well as a lateral shape is a result of the advanced fabrication process in which the shape formation is mostly affected by etching. Figure 2 shows a comparison of experimental I_D-V_G characteristics of the 22 nm gate length GAA-NW FET versus the simulation results provided by VENDES SCH-MC. The drain bias is 1.0 V. The empirical doping values were not included in [14], so they were reversed engineered following the methodology described in [16]. Note that, the MC device simulations in VENDES are able to accurately reproduce the experimental results in all the active regions of the device, except for at a very low gate bias of 0.0 V, where the MC statistical noise is too high. At the very low gate bias, the DD device simulations are typically used.

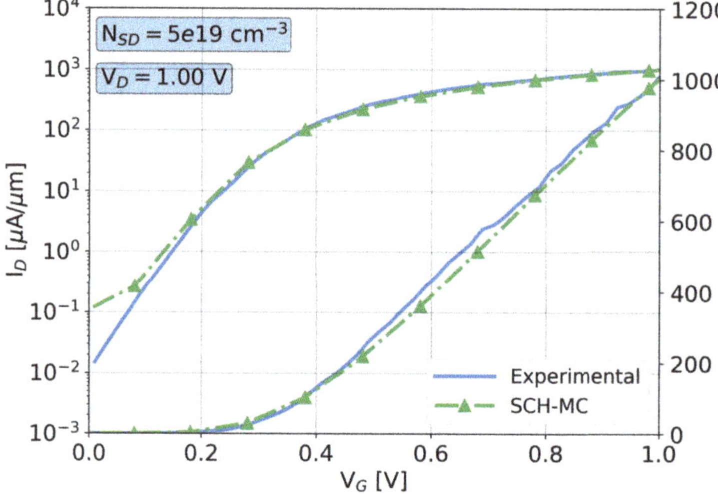

Figure 2. I_D-V_G characteristics for a 22 nm gate length GAA-NW FET (gate-all-around nanowire field effect transistor) at a drain bias of 1.0 V comparing experimental results from [14] against Schrödinger equation corrected Monte Carlo (SCH-MC) simulations from VENDES. The maximum source/drain Gaussian doping has been set to 5×10^{19} cm^{-3}.

The main dimensions and doping values used to model the 10 nm gate length Si GAA-NW FET are summarised in Table 1. The gate work-function (WF) was set to 4.4 eV. For this device, Figure 3 shows the I_D-V_G characteristics at a high drain bias of 0.7 V in both linear and logarithmic scales for DG-DD, SCH-DD and SCH-MC simulations. Note that SCH-MC simulations are calibration free, whereas the DG and DD models need to be properly fitted (see the main calibration parameters in Table 1) in order to achieve the agreement shown in Figure 3. The main figures of merit (FoM) that characterise the I_D-V_G characteristics are shown in Table 1.

FoMPy [36,37] is a python-based open source post-processing tool implemented in VENDES (see Figure 1) that automatically extracts the main FoMs of a *I*-*V* characteristics. This tool is very useful when performing statistical studies, where a large ensemble of devices needs to be analysed. In this work, the threshold voltage (V_T) has been obtained using the second derivation method, the off-current is obtained at a 0.0 V gate voltage, and the on-current has been extracted at a gate bias equal to $V_T + V_{DD}$, being V_{DD} the supply voltage (set to 0.7 V). The analysed device has a low off-current of 0.03 μA/μm, acceptable for applications in mobile low power devices with a long battery life, and an on-current of 1770 μA/μm, that has been achieved by increasing the maximum S/D doping from 5×10^{19} cm^{-3}, used in the 22 nm gate length experimental device, to 10^{20} cm^{-3}. This increase in the doping has allowed to raise the device on-current by 40% (as previously shown in [38]), at the cost of a slight deterioration in the sub-threshold slope (SS). The device SS is 71 mV/dec, not far from the ideal limit of 60 mV/dec. Therefore, the device doping is one of the key design parameters that needs to be considered when designing a device for a specific application. For transistors aimed at high performance (HP), standard performance (SP) and low power (LP) applications, the I_{ON}/I_{OFF} ratio is also a key parameter because it provides a global characterisation of the device operation. The observed I_{ON}/I_{OFF} (6.63×10^4) is 27% larger than that of a similar gate length Si FinFET [7] of 10.7 nm.

Table 1. Device dimensions, doping values, main FoMs and calibration parameters for the 10 nm gate length Si GAA-NW FET.

Dimensions	Gate length (L_G) (nm)	10.0
	Source and drain (S/D) length ($L_{S/D}$) (nm)	14.0
	Channel width (W_{CH}) (nm)	5.70
	Channel height (H_{CH}) (nm)	7.17
	Equivalent oxide thickness (EOT) (nm)	0.80
Doping values	S/D *n*-type doping (N_{SD}) (cm^{-3})	10^{20}
	S/D doping lateral straggle (σ_x)	3.23
	S/D doping lateral peak (x_{max}) (nm)	7.80
Figures of merit	Subthreshold slope (SS) (mV/dec)	71.0
	Threshold voltage (V_T) (V)	0.250
	Off-current (I_{OFF}) (μA/μm)	0.027
	On-current (I_{ON}) (μA/μm)	1770
	I_{ON}/I_{OFF} ratio	6.63×10^4
Calibration parameters	Saturation velocity (v_{sat}) (cm/s)	1.30×10^7
	Perpendicular critical electric field (E_{CN})(V/cm)	9.95×10^5
	DG electron mass in the transport direction (m_x)(m_0)	0.50
	DG electron masses in the confinement direction ($m_{y,z}$)(m_0)	0.10

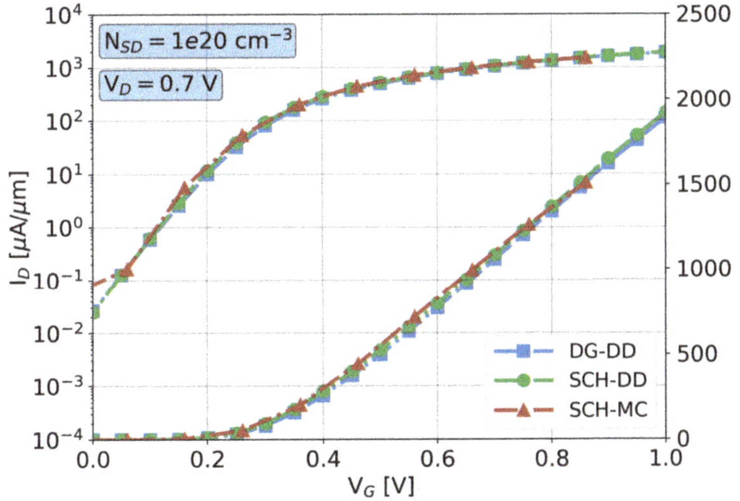

Figure 3. I_D-V_G characteristics for a 10 nm gate length GAA-NW FET at a drain bias of 0.7 V comparing simulations results from density-gradient quantum-corrected drift-diffusion simulations (DG-DD), Schrödinger quantum-corrected drift-diffusion simulations (SCH-DD) and SCH-MC. The maximum source/drain Gaussian doping has been set to 10^{20} cm^{-3}.

3.2. Variability Models

Several sources of intrinsic device variability are considered in VENDES: i) Metal Grain Granularity (MGG) [39], ii) line edge roughness (LER) [16], iii) gate edge roughness (GER) [7] and iv) random discrete dopants (RD) [40]. These variability sources, together with oxide thickness variations (OTV) and interface trap charges (ITC), were shown to affect FinFETs and GAA-NW FETs the most [5,41].

The MGG is modelled by altering the work-function of the device gate so it matches metal grain distributions either observed empirically via KPFM [42], or generated using the Voronoi approach, where the experimental shapes and values of different grain orientations are mimicked [43]. Figure 4a shows an example of a Voronoi TiN metal profile applied to the device gate. The WF values are 4.4 eV and 4.6 eV and their respective probabilities of occurrence 40% and 60%. The average grain size is 5 nm.

RD are introduced in the *n*-type doped S/D regions using a rejection technique from the doping profile (shown in Table 1) of the ideal device. Initially, dopants with their associated charge are distributed on an atomistic grid defined by the location of the atoms. Then, this charge is mapped to the device tetrahedral mesh using the cloud-in-cell technique, in order to generate an atomistic electron density distribution [44], as shown in Figure 4b.

GER and LER are modelled similarly, the device gate (in case of the GER) or the edge of the nanowire (in case of the LER) are deformed according to the shape of a given roughness profile created via the Fourier synthesis method [45]. Two parameters are used to characterise these deformations: i) the root mean square (RMS) height, that sets the amplitude of the roughness, and ii) the correlation length (CL), that accounts for the spatial correlation between the deformations in the different points of the device. Figure 4c,d show examples of devices affected by the GER (with a CL = 11 nm) and the LER (with a CL = 20 nm), respectively. In both cases the RMS height is 0.8 nm.

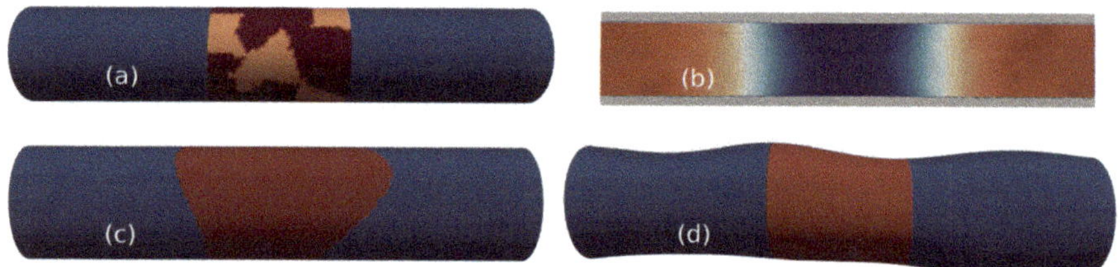

Figure 4. Examples of different sources of variability applied to the 10 nm gate length GAA-NW FET: (**a**) TiN metal profile (with work-function values of 4.4 eV and 4.6 eV) applied to the device gate leading to metal grain granularity (MGG) variations, (**b**) effect of the random dopants (RD) present in the source/drain regions of the device on the device electron concentration, (**c**) device gate affected by gate edge roughness (GER) and (**d**) device body under line edge roughness (LER) variations.

For each variability source, ensembles of 300 device configurations were created and simulated at a high drain bias of 0.7 V. Figure 5 shows the impact of the aforementioned sources on the 10 nm GAA-NW FET threshold voltage variability. The statistical sum of the four sources of variability (COMB) has also been included as comparison. Results show that GAA-NW FETs are heavily influenced by the LER variability in the sub-threshold region, with σV_T values 1.4 and 2.0 times larger than those of the MGG and the RD, respectively. The GER is the least influential source of variability having its σV_T a 86% lower than that of the LER. Note that the combination of the four sources of variability leads to a V_T standard deviation of 55.5 mV, a value 85% larger than the one observed (σV_T = 30 mV) in a similar gate length Si FinFET [7] of 10.7 nm.

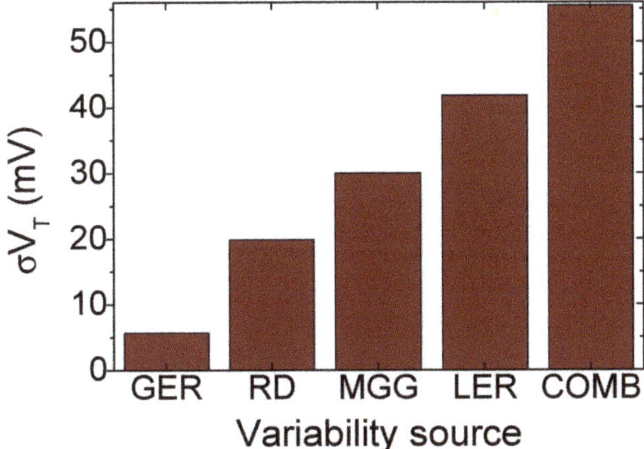

Figure 5. Comparison of the threshold voltage standard deviation due to four different variability sources (GER, RD, MGG and LER) and their combined effect (COMB) in the 10 nm GAA-NW FET.

Variability studies are highly computational demanding because they require the simulation of hundreds or thousands of device configurations in order to obtain results with statistical significance. Table 2 shows, for the different simulation methodologies implemented in VENDES, the total times for the solution of one I_D-V_G bias point at a high drain bias of 0.7 V on a single core for two different CPUs. Note that the simulation time for a SCH-MC simulation is around 70 times longer than a quantum-corrected DD study. For that reason, VENDES performs sub-threshold region variability studies (see the flowchart in Figure 1) using either DG or SCH-DD simulations. The reason for this is twofold: i) in the sub-threshold, the electrostatics dominate and quantum-corrected DD simulations are able to provide accurate results and, ii) MC results can be extremely noisy at very low gate biases and lead to incorrect off-current or sub-threshold slope values. However, in the on-region regime, VENDES performs the variability studies via the SCH-MC simulations since the DD approach is unable to capture a non-equilibrium carrier transport even if it is properly calibrated, leading to large over- or under-estimation of the variability [38]. However, it is important to remark (as seen in Figure 3), that both SCH-DD and MC-DD simulations match perfectly at the threshold.

In the quest for the reduction of the computational time, several alternatives have been investigated: i) the parallelisation of the simulation code using a message passing interface (MPI), explained in detail in [46], to take advantage of increasingly available computational infrastructures, such as clusters and supercomputers and, ii) the development of a methodology to predict the impact of the variability sources. A sequential simulation of one I_D-V_G bias point using the SCH-DD method is 960 s, see Table 2 for runs using the Intel(R) Xeon(R) CPU. When using a parallel version of the code with 2 and 4 processors, this time is reduced to 613 s (78 % parallel efficiency) and 363 s (66 % parallel efficiency), respectively.

Table 2. Total time for the solution of one gate bias point as a function on the simulation method. Results have been obtained on a single core for two different CPUs: an AMD Opteron 6262HE @ 1.60 GHz and a Intel(R) Xeon(R) E5-2643 v2 @ 3.50 GHz.

Simulation Method	AMD Time (hh:mm)	Intel Time (hh:mm)
DD	00:28	00:08
DG-DD	01:02	00:20
SCH-DD	00:55	00:16
SCH-MC	70:00	35:00

On the other hand, the Fluctuation Sensitivity Map (FSM) approach [47,48] is a methodology that we developed to predict the impact of the variability sources. This post-processing tool (see the VENDES flowchart in Figure 1) is based on the creation of a map that provides information of the sensitivity of the different regions of a device to a particular source of variability. Once the map is created, it can be used to predict the statistical variability, with a very small error, under different input parameters as shown in Ref. [47]. In a typical variability study, we simulate at least 300 device configurations per variability source and characteristic parameter. This characteristic parameter can be the grain size in the MGG study, or the CL and RMS in a study of LER and GER. For instance, a full MGG variability study will require simulations of several grain sizes (a minimum of three). Therefore, the total computational cost of this full study using the SCH-DD or SCH-MC methods will be 300 h and 31500 h, respectively (for the Intel(R) Xeon(R) CPU in the sequential case). These times can be reduced by 66% using the FSM because once one set of 300 device configurations is simulated and used to create the FSM (which can take up to 2 min), the map can be used to predict the variability results for the remaining grain sizes without any further statistical computations.

The FSM can be also used as an assistance in the design of variability-resistant device architectures since it pinpoints to parts of the device the manufacturers should concentrate their efforts on. Figure 6 shows an example of the on-current FSM obtained when a single LER deformation is applied to a specific location of the device, narrowing its width. Using this synthetic deformation, it is possible to sweep all the locations along the device, measuring the changes in the FoM that enable us to the spatial sensitivity to LER variations. Note that, in Figure 6 (bottom), a negative (positive) sensitivity indicates an increase (decrease) in the on-current. The effect of the synthetic deformation on the on-current depends on its position along the transistor. Any change in the NW width happening near the source-gate junction will heavily impact the device on-current whereas if the deformation occurs near the source or drain ends, its impact on the on-current will be minimal.

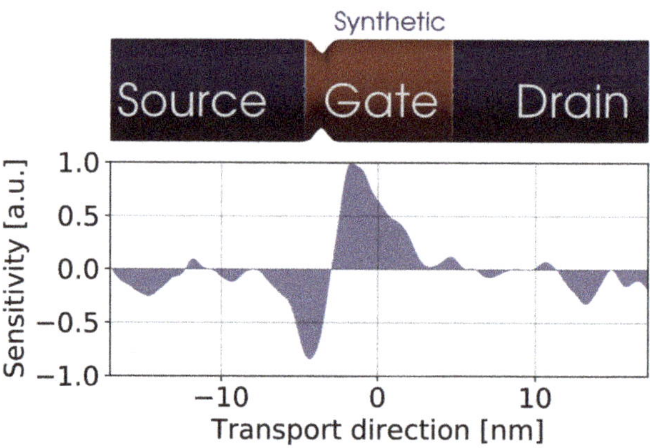

Figure 6. (**Top**) Schematic of a 10 nm gate length GAA-NW FET affected by a localised LER deformation. (**Bottom**) 1D on-current FSM generated from the simulation of 100 localised LER profiles swept along the device channel at a 0.7 V drain bias.

Similarly, the FSM can be applied to other sources of variability, like the MGG, taking into account that the generated fluctuation map will now be two-dimensional (2D), in order to characterise the whole device gate. Figure 7a shows a scheme of the device that has a fixed WF value of 4.6 eV in all the gate except from a narrow strip in which the WF is 4.4 eV. This narrow strip is swept along the gate and for each position, the device configuration was simulated at a high drain bias of 0.7 V and the corresponding on-current was extracted. The resulting 2D on-current FSM due to the MGG is shown in Figure 7b. This map allows us to establish that any variation in the WF located between X = 0.0 nm (middle of the gate) and $X = -2.5$ nm, will have the largest impact on the device performance. However, the GAA-NW FET will be practically insensitive to WF variations happening in the proximity of the source ($X = -5.0$ nm) or the drain ($X = 5.0$ nm) junctions.

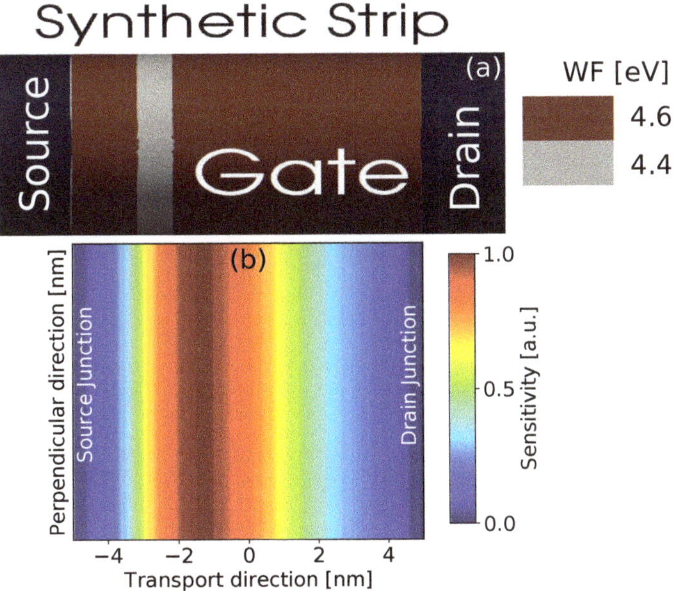

Figure 7. (**a**) Schematic of a 10 nm gate length GAA-NW FET affected by a synthetic MGG profile. The WF is 4.6 eV in all the gate except from a narrow strip (0.1 nm wide) with a WF 4.4 eV. (**b**) 2D on-current FSM generated from the simulation of 100 synthetic gate profiles swept along the device at a 0.7 V drain bias.

4. Conclusions

VENDES, an in-house-built 3D multi-method device simulator, has been used to characterise the performance and resistance to variability of a 10 nm gate length Si GAA-NW FET scaled down from an experimental transistor [14] following ITRS guidelines [35]. The off-current of the device is 0.03 µA/µm, and the on-current of 1770 µA/µm, delivering the I_{ON}/I_{OFF} ratio of 6.63×10^4. The device SS is 71 mV/dec, not far from the ideal limit of 60 mV/dec. σV_T due to the statistical combination of LER, GER, MGG and RD is 55.5 mV, a value significantly larger than that of a similar gate length Si FinFET of 10.7 nm (30 mV). This larger threshold voltage variability indicates that the variability effects may be another limiting factor for the adoption of the GAA-NW FETs in the future technological nodes.

Finally, the FSM allowed us to determine which regions of the device are the most sensitive to the LER and MGG variations and influence the device characteristics the most. In the case of the LER, the changes in the device width occurring near the source-gate junction will heavily impact the device on-current whereas any deformation happening near the source or the drain, will have a negligible influence on the on-current. In the case of the MGG, the most sensitive region of the device is localised between the middle of the gate and the proximity of the gate-source junction. The information provided by the FSM can be very useful as an aid for the creation of fluctuation-resistant device architectures.

Author Contributions: Research conceptualization, N.S. and A.G.-L.; methodology, G.I., N.S. and A.G.-L.; software, D.N., G.E, K.K. and G.I.; validation, D.N., G.E, G.I., N.S. and A.G.-L; writing—original draft preparation, N.S. and K.K.; supervision, K.K. and A.G.-L.

Funding: This research was supported in part by the Spanish Government under the projects TIN2013-41129-P, TIN2016-76373-P and RYC-2017-23312, by Xunta de Galicia and FEDER funds (GRC 2014/008) and by the Consellería de Cultura, Educación e Ordenación Universitaria (accreditation 2016-2019, ED431G/08).

Acknowledgments: The authors thank Centro de Supercomputación de Galicia (CESGA) for the computer resources provided.

Conflicts of Interest: The authors declare no conflict of interest.

Abbreviations

The following abbreviations are used in this manuscript:

DD	Drift diffusion
DG	Density gradient
FE	Finite-element
FinFET	Fin field effect transistor
FSM	Fluctuation sensitivity map
GAA-NW FET	Gate-all-around nanowire field effect transistor
GER	Gate edge roughness
KPFM	Kelvin probe force microscopy
LER	Line edge roughness
MC	Monte Carlo
MGG	Metal grain granularity
RD	Random discrete dopants

References

1. Badami, O.; Driussi, F.; Palestri, P.; Selmi, L.; Esseni, D. Performance comparison for FinFETs, nanowire and stacked nanowires FETs: Focus on the influence of surface roughness and thermal effects. In Proceedings of the 2017 IEEE International Electron Devices Meeting (IEDM), San Francisco, CA, USA, 2–6 December 2017; pp. 13.2.1–13.2.4. doi:10.1109/IEDM.2017.8268382. [CrossRef]
2. Yoon, J.S.; Rim, T.; Kim, J.; Meyyappan, M.; Baek, C.K.; Jeong, Y.H. Vertical gate-all-around junctionless nanowire transistors with asymmetric diameters and underlap lengths. *J. Appl. Phys.* **2014**, *105*, 102105. doi:10.1063/1.4895030. [CrossRef]
3. Mikolajick, T.; Heinzig, A.; Trommer, J.; Pregl, S.; Grube, M.; Cuniberti, G.; Weber, W. Silicon nanowires—A versatile technology platform. *Phys. Status Solidi Rapid Res. Lett.* **2013**, *7*, 793–799. doi:10.1002/pssr.201307247. [CrossRef]
4. IEEE International Roadmap for Devices and Systems (IRDS), More Moore. 2017. Available online: https://irds.ieee.org/roadmap-2017 (accessed on 24 July 2019).
5. Wang, X.; Brown, A.R.; Cheng, B.; Asenov, A. Statistical variability and reliability in nanoscale FinFETs. In Proceedings of the 2011 International Electron Devices Meeting, Washington, DC, USA, 5–7 December 2011; pp. 5.4.1–5.4.4. doi:0.1109/IEDM.2011.6131494. [CrossRef]
6. Nagy, D.; Indalecio, G.; García-Loureiro, A.J.; Elmessary, M.A.; Kalna, K.; Seoane, N. FinFET Versus Gate-All-Around Nanowire FET: Performance, Scaling, and Variability. *IEEE J. Electron Devices Soc.* **2018**, *6*, 332–340. doi:10.1109/JEDS.2018.2804383. [CrossRef]
7. Espiñeira, G.; Nagy, D.; Indalecio, G.; García-Loureiro, A.J.; Kalna, K.; Seoane, N. Impact of Gate Edge Roughness Variability on FinFET and Gate-All-Around Nanowire FET. *IEEE Electron Device Lett.* **2019**, *40*, 510–513. doi:10.1109/LED.2019.2900494. [CrossRef]
8. Vasileska, D.; Goodnick, S.M.; Klimeck, G. *Computational Electronics: Semiclassical and Quantum Device Modeling and Simulation;* CRC Press: Boca Raton, USA. 2010.
9. Asenov, A.; Cheng, B.; Wang, X.; Brown, A.R.; Millar, C.; Alexander, C.; Amoroso, S.M.; Kuang, J.B.; Nassif, S.R. Variability Aware Simulation Based Design—Technology Cooptimization (DTCO) Flow in 14 nm FinFET/SRAM Cooptimization. *IEEE Trans. Electron Devices* **2015**, *62*, 1682–1690. doi:10.1109/TED.2014.2363117. [CrossRef]
10. Selberherr, S. *Simulation of Semiconductor Devices and Processes;* Springer-Verlag: Wien, Vienna; New York, NY, USA, 1993.

11. Fang, J.; Vandenberghe, W.; Fu, B.; Fischetti, M. Pseudopotential-based electron quantum transport: Theoretical formulation and application to nanometer-scale silicon nanowire transistors. *J. Appl. Phys.* **2016**, *119*, 035701. [CrossRef]
12. Datta, S. Nanoscale device modeling: The Green's function method. *Superlattices Microstruct.* **2000**, *28*, 253–278. doi:10.1006/spmi.2000.0920. [CrossRef]
13. Luisier, M.; Klimeck, G. Atomistic full-band simulations of silicon nanowire transistors: Effects of electron-phonon scattering. *Phys. Rev. B* **2009**, *80*, 155430. doi:10.1103/PhysRevB.80.155430. [CrossRef]
14. Bangsaruntip, S.; Balakrishnan, K.; Cheng, S.L.; Chang, J.; Brink, M.; Lauer, I.; Bruce, R.L.; Engelmann, S.U.; Pyzyna, A.; Cohen, G.M.; et al. Density scaling with gate-all-around silicon nanowire MOSFETs for the 10 nm node and beyond. In Proceedings of the IEEE Electron Devices Meeting (IEDM), Washington, DC, USA, 9–11 December 2013; pp. 20.2.1–20.2.4. doi:10.1109/IEDM.2013.6724667. [CrossRef]
15. Geuzaine, C.; Remacle, J.F. Gmsh: A three-dimensional finite element mesh generator with built-in pre- and post-processing facilities. *Int. J. Numer. Meth. Eng.* **2009**, *79*, 1309–1331. doi:10.1002/nme.2579. [CrossRef]
16. Elmessary, M.A.; Nagy, D.; Aldegunde, M.; Seoane, N.; Indalecio, G.; Lindberg, J.; Dettmer, W.; Perić, D.; García-Loureiro, A.J.; Kalna, K. Scaling/LER study of Si GAA nanowire FET using 3D finite element Monte Carlo simulations. *Solid-State Electron.* **2017**, *128*, 17 – 24. Extended papers selected from EUROSOI-ULIS 2016, doi:10.1016/j.sse.2016.10.018. [CrossRef]
17. Garcia-Loureiro, A.J.; Seoane, N.; Aldegunde, M.; Valin, R.; Asenov, A.; Martinez, A.; Kalna, K. Implementation of the Density Gradient Quantum Corrections for 3-D Simulations of Multigate Nanoscaled Transistors. *IEEE Trans. Comput.-Aided Des. Integr. Circuits Syst.* **2011**, *30*, 841–851. doi:10.1109/TCAD.2011.2107990. [CrossRef]
18. Ancona, M.G.; Yu, Z.; Dutton, R.W.; Voorde, P.J.V.; Cao, M.; Vook, D. Density-Gradient Analysis of MOS Tunneling. *IEEE Trans. Electron Devices* **2000**, *47*, 2310–2319. doi:10.1109/16.887013. [CrossRef]
19. Asenov, A.; Watling, J.R.; Brown, A.R.; Ferry, D.K. The Use of Quantum Potentials for Confinement and Tunnelling in Semiconductor Devices. *J. Comput. Electron.* **2002**, *1*, 503–513. doi:10.1023/A:1022905508032. [CrossRef]
20. Seoane, N.; Indalecio, G.; Comesana, E.; Aldegunde, M.; García-Loureiro, A.J.; Kalna, K. Random Dopant, Line-Edge Roughness, and Gate Workfunction Variability in a Nano InGaAs FinFET. *IEEE Trans. Electron Devices* **2014**, *61*, 466–472. doi:10.1109/TED.2013.2294213. [CrossRef]
21. Kovac, U.; Alexander, C.; Roy, G.; Riddet, C.; Cheng, B.; Asenov, A. Hierarchical Simulation of Statistical Variability: From 3-D MC With ab initio Ionized Impurity Scattering to Statistical Compact Models. *IEEE Trans. Electron Devices* **2010**, *57*, 2418–2426. doi:10.1109/TED.2010.2062517. [CrossRef]
22. Winstead, B.; Ravaioli, U. A quantum correction based on Schrodinger equation applied to Monte Carlo device simulation. *IEEE Trans. Electron Devices* **2003**, *50*, 440–446. doi:10.1109/TED.2003.809431. [CrossRef]
23. Lindberg, J.; Aldegunde, M.; Nagy, D.; Dettmer, W.G.; Kalna, K.; García-Loureiro, A.J.; Perić, D. Quantum Corrections Based on the 2-D Schrödinger Equation for 3-D Finite Element Monte Carlo Simulations of Nanoscaled FinFETs. *IEEE Trans. Electron Devices* **2014**, *61*, 423–429. doi:10.1109/TED.2013.2296209. [CrossRef]
24. Elmessary, M.A.; Nagy, D.; Aldegunde, M.; Lindberg, J.; Dettmer, W.G.; Perić, D.; García-Loureiro, A.J.; Kalna, K. Anisotropic Quantum Corrections for 3-D Finite-Element Monte Carlo Simulations of Nanoscale Multigate Transistors. *IEEE Trans. Electron Devices* **2016**, *63*, 933–939. doi:10.1109/TED.2016.2519822. [CrossRef]
25. Assad, F.; Banoo, K.; Lundstrom, M. The drift-diffusion equation revisited. *Solid-State Electron.* **1997**, *42*, 283–295. doi:10.1016/S0038-1101(97)00263-3. [CrossRef]
26. Caughey, D.M.; Thomas, R.E. Carrier Mobilities in Silicon Empirically Related to Doping and Field. *Proc. IEEE* **1967**, *55*, 2192–2193. doi:10.1109/PROC.1967.6123. [CrossRef]
27. Yamaguchi, K. Field-dependent mobility model for two-dimensional numerical analysis of MOSFET's. *IEEE Trans. Electron Devices* **1979**, *26*, 1068–1074. doi:10.1109/T-ED.1979.19547. [CrossRef]
28. Jacoboni, C.; Lugli, P. *The Monte Carlo Method for Semiconductor Device Simulation;* Computational Microelectronics; Springer: Vienna, Austria. 2012.
29. Aldegunde, M.; García-Loureiro, A.J.; Kalna, K. 3D Finite Element Monte Carlo Simulations of Multigate Nanoscale Transistors. *IEEE Trans. Electron Devices* **2013**, *60*, 1561–1567. doi:10.1109/TED.2013.2253465. [CrossRef]

30. Tomizawa, K. *Numerical Simulation of Submicron Semiconductor Devices*; Artech House Materials Science Library, Artech House: Boston, USA. 1993.
31. Ridley, B.K. Reconciliation of the Conwell-Weisskopf and Brooks-Herring formulae for charged-impurity scattering in semiconductors: Third-body interference. *J. Phys. C Solid State Phys.* **1977**, *10*, 1589. [CrossRef]
32. de Roer, T.G.V.; Widdershoven, F.P. Ionized impurity scattering in Monte Carlo calculations. *J. Appl. Phys.* **1986**, *59*, 813–815. doi:10.1063/1.336603. [CrossRef]
33. Ferry, D. *Semiconductor Transport*; Taylor & Francis: London, United Kingdom. 2000.
34. Islam, A.; Kalna, K. Monte Carlo simulations of mobility in doped GaAs using self-consistent Fermi–Dirac statistics. *Semicond. Sci. Technol.* **2012**, *26*, 039501. [CrossRef]
35. International Technology Roadmap for Semiconductors (ITRS). 2016. Available online: http://www.itrs2.net/ (accessed on 24 July 2019).
36. Espiñeira, G.; Seoane, N.; Nagy, D.; Indalecio, G.; García-Loureiro, A.J. FoMPy: A figure of merit extraction tool for semiconductor device simulations. In Proceedings of the 2018 Joint International EUROSOI Workshop and International Conference on Ultimate Integration on Silicon (EUROSOI-ULIS), Granada, Spain, 19–21 March 2018. doi:10.1109/ULIS.2018.8354752. [CrossRef]
37. Espiñeira, G.; Nagy, D.; García-Loureiro, A.J.; Seoane, N.; Indalecio, G. Impact of threshold voltage extraction methods on semiconductor device variability. *Solid-State Electron.* **2019**, *159*, 165–170. doi:10.1016/j.sse.2019.03.055. [CrossRef]
38. Nagy, D.; Indalecio, G.; García-Loureiro, A.J.; Espiñeira, G.; Elmessary, M.A.; Kalna, K.; Seoane, N. Drift-Diffusion Versus Monte Carlo Simulated ON-Current Variability in Nanowire FETs. *IEEE Access* **2019**, *7*, 12790–12797. doi:10.1109/ACCESS.2019.2892592. [CrossRef]
39. Indalecio, G.; Aldegunde, M.; Seoane, N.; Kalna, K.; García-Loureiro, A.J. Statistical study of the influence of LER and MGG in SOI MOSFET. *Semicond. Sci. Technol.* **2014**, *29*, 045005. [CrossRef]
40. Seoane, N.; Indalecio, G.; Nagy, D.; Kalna, K.; García-Loureiro, A.J. Impact of Cross-Sectional Shape on 10-nm Gate Length InGaAs FinFET Performance and Variability. *IEEE Trans. Electron Devices* **2018**, *65*, 456–462. doi:10.1109/TED.2017.2785325. [CrossRef]
41. Wang, R.; Zhuge, J.; Huang, R.; Yu, T.; Zou, J.; Kim, D.W.; Park, D.; Wang, Y. Investigation on Variability in Metal-Gate Si Nanowire MOSFETs: Analysis of Variation Sources and Experimental Characterization. *IEEE Trans. Electron Devices* **2011**, *58*, 2317–2325. doi:10.1109/TED.2011.2115246. [CrossRef]
42. Ruiz, A.; Seoane, N.; Claramunt, S.; García-Loureiro, A.; Porti, M.; Couso, C.; Martin-Martinez, J.; Nafria, M. Workfunction fluctuations in polycrystalline TiN observed with KPFM and their impact on MOSFETs variability. *Appl. Phy. Lett.* **2019**, *114*, 093502. doi:10.1063/1.5090855. [CrossRef]
43. Indalecio, G.; García-Loureiro, A.J.; Iglesias, N.S.; Kalna, K. Study of Metal-Gate Work-Function Variation Using Voronoi Cells: Comparison of Rayleigh and Gamma Distributions. *IEEE Trans. Electron Devices* **2016**, *63*, 2625–2628. doi:10.1109/TED.2016.2556749. [CrossRef]
44. Asenov, A.; Brown, A.R.; Roy, G.; Cheng, B.; Alexander, C.; Riddet, C.; Kovac, U.; Martinez, A.; Seoane, N.; Roy, S. Simulation of statistical variability in nano-CMOS transistors using drift-diffusion, Monte Carlo and non-equilibrium Green's function techniques. *J. Comput. Electron.* **2009**, *8*, 349–373. doi:10.1007/s10825-009-0292-0. [CrossRef]
45. Asenov, A.; Kaya, S.; Brown, A.R. Intrinsic parameter fluctuations in decananometer MOSFETs introduced by gate line edge roughness. *IEEE Trans. Electron Devices* **2003**, *50*, 1254–1260. doi:10.1109/TED.2003.813457. [CrossRef]
46. Seoane, N.; García-Loureiro, A.J.; Aldegund, M. Optimisation of linear systems for 3D parallel simulation of semiconductor devices: Application to statistical studies. *Int. J. Numer. Model. Electron. Netw. Devices Fields* **2009**, *22*, 235–258. doi:10.1002/jnm.695. [CrossRef]

47. Indalecio, G.; Seoane, N.; Kalna, K.; García-Loureiro, A.J. Fluctuation Sensitivity Map: A Novel Technique to Characterise and Predict Device Behaviour Under Metal Grain Work-Function Variability Effects. *IEEE Trans. Electron Devices* **2017**, *64*, 1695–1701. doi:10.1109/TED.2017.2670060. [CrossRef]
48. Indalecio, G.; García-Loureiro, A.J.; Elmessary, M.A.; Kalna, K.; Seoane, N. Spatial Sensitivity of Silicon GAA Nanowire FETs Under Line Edge Roughness Variations. *IEEE J. Electron Devices Soc.* **2018**, *6*, 601–610. doi:10.1109/JEDS.2018.2828504. [CrossRef]

Article

Characteristic Fluctuations of Dynamic Power Delay Induced by Random Nanosized Titanium Nitride Grains and the Aspect Ratio Effect of Gate-All-Around Nanowire CMOS Devices and Circuits

Yiming Li *, Chieh-Yang Chen, Min-Hui Chuang and Pei-Jung Chao

Parallel and Scientific Computing Laboratory, Department of Electrical and Computer Engineering, National Chiao Tung University, Hsinchu 300, Taiwan; cychen@mail.ymlab.org (C.-Y.C.); mhchuang@mail.ymlab.org (M.-H.C.); pjchao@mail.ymlab.org (P.-J.C.)

* Correspondence: ymli@faculty.nctu.edu.tw; Tel.: +886-3-5712121 (ext. 52974)

Received: 1 January 2019; Accepted: 5 May 2019; Published: 8 May 2019

Abstract: In this study, we investigate direct current (DC)/alternating current (AC) characteristic variability induced by work function fluctuation (WKF) with respect to different nanosized metal grains and the variation of aspect ratios (ARs) of channel cross-sections on a 10 nm gate gate-all-around (GAA) nanowire (NW) metal–oxide–semiconductor field-effect transistor (MOSFET) device. The associated timing and power fluctuations of the GAA NW complementary metal–oxide–semiconductor (CMOS) circuits are further estimated and analyzed simultaneously. The experimentally validated device and circuit simulation running on a parallel computing system are intensively performed while considering the effects of WKF and various ARs to access the device's nominal and fluctuated characteristics. To provide the best accuracy of simulation, we herein calibrate the simulation results and experimental data by adjusting the fitting parameters of the mobility model. Transfer characteristics, dynamic timing, and power consumption of the tested circuit are calculated using a mixed device–circuit simulation technique. The timing fluctuation mainly follows the trend of the variation of threshold voltage. The fluctuation terms of power consumption comprising static, short-circuit, and dynamic powers are governed by the trend that the larger the grain size, the larger the fluctuation.

Keywords: DC and AC characteristic fluctuations; gate-all-around; nanowire; MOSFETs; work function fluctuation; aspect ratio of channel cross-section; timing fluctuation; noise margin fluctuation; power fluctuation; CMOS circuit; statistical device simulation

---

1. Introduction

The dimension of effective devices has shrunk to a sub-22 nanometer scale, and due to this, we are facing even more serious characteristic variability problems [1–7]. High-κ/metal gate (HKMG) technology has been recognized as a solution to solve intrinsic fluctuation, but the crystal orientation of nanosized metal grain is uncontrollable during the growth step under high temperatures [8,9]. Values of uncertain orientation-dependent work functions (WKs) of gate material causes WK fluctuation (WKF). Many studies have surveyed WKF for different devices [3,10–16], and some have further discussed the distribution of metal grains on planar metal–oxide–semiconductor field-effect transistors (MOSFETs) [17,18]. However, seldom do these studies put emphasis on gate-all-around (GAA) nanowire (NW) MOSFET devices. As a result, in this study, we will focus on estimating the impact of WKF on the electrical characteristics of GAA NW MOSFET devices and its implication for the dynamic property of a complementary metal–oxide–semiconductor (CMOS) circuit. In order to analyze WKF, we apply the newly developed localized WKF (LWKF) method [18]. The averaged work function method (AWKF) [19] was reported to estimate the entire value of WKs, but the process of averaging cannot be used to estimate the effect of local WKs on device characteristics. The physically-sound LWKF method is an effective technique that can determine the random number and random location effects, as well as the physical phenomena of the localization of nanosized metal grains, and it does not underestimate the effect of WKF [20]. Hence, we adopt the LWKF method to explore the WKF effect. According to the properties of the metal material, TiN has two different orientations: <200> and <111>, with 60% and 40% generated probabilities [19,21,22]. In addition to WKF, the limitation of process fabrication may lead to geometrical variations in channel cross-sections and affect the operations of devices. Due to this, perfectly round-shaped channel GAA NW MOSFET devices are difficult to manufacture. The different aspect ratio (AR) of channel radius results in a different shape of channel cross-section—an elliptical shape instead of the ideal round shape [23–25]. Therefore, we will discuss the electrical characteristics of the explored devices with different ARs. Additionally, we will simulate the combination effect of WKF and the variation of AR. WKF-induced circuit variations, such as timing and power fluctuations [26–28], seriously affect the dynamic property of GAA NW CMOS circuits. Most previous research only focused on the DC characteristics of N-type planar or fin-typed MOSFET devices when considering the aforementioned variability [10,28–30]. Various fluctuations of circuit characteristics, such as noise margin (NM), timing, and power consumption are also important to research, but the variability of GAA NW CMOS circuits has not been clearly studied yet. To comprehensively explore the aforementioned issues for 10 nm gate GAA NW MOSFETs and CMOS circuits induced by WKF and different ARs, we extend an experimentally-calibrated three-dimensional (3D) quantum-mechanically-corrected device and circuit simulation [1,18,26–29]. The engineering findings of this study indicate that falling time (t_f) is lower than rising time (t_r) owing to the relatively larger driving capability of the N-type device. Along with the increasing grain number of higher WKs, the high-to-low delay time (t_{HL}) and the low level of noise margin (NM_L) become higher, while the low-to-high delay time (t_{LH}) and the high level of noise margin (NM_H) decrease. All power consumption terms follow the trend that the larger the grain size, the larger the fluctuation.

This paper is organized as follows: In Section 2, we describe the statistical device simulation techniques of WKF and AR for the GAA NW MOSFETs and CMOS circuits. In Section 3, we discuss and analyze the simulation results of WKF combined with the variation of AR on GAA NW MOSFETs and CMOS circuits. Finally, we draw the conclusions of this study and suggest future work.

2. Statistical LWKF and AR Simulation Techniques

In this work, we extended the statistical device simulation technique [3,31] to analyze WKF and different ARs of GAA NW CMOS circuits. Figure 1a shows the device setting parameters, the device characteristics, and the achieved nominal values of the short-channel effect (SCE) of the studied N-/P-type devices. To conduct the simulation and to estimate the impacts of WKF, we used the LWKF method for statistical device simulation, which is illustrated in Figure 1b–e in detail. We used TiN as the metal gate material, which includes two different orientations: <200> and <111> with the associated 60% and 40% probabilities. The related parameters are shown in Figure 1b. To calibrate the magnitude of threshold voltage (V_{th}) to 280 mV, we used the WK-tuning techniques in which the metal gate is doped by hydrogen plasma/fluorine ion implantation, as this was found by K. Han et al. [32,33] to achieve different WK values. Thus, the corresponding WKs are 4.6 and 4.84 eV and 4.4 and 4.64 eV, respectively, for the N-/P-type devices. First, to carry out the WKF simulation, we partitioned the TiN metal gate of the GAA NW MOSFET devices into many sub-regions according to grain size. Second, Figure 1c shows a histogram plot of the number of high WKs, which were generated according to Gaussian distribution. Then, the high and low WKs were randomly assigned and mapped onto the sub-region of the gate region of device, as shown in Figure 1d. Finally, we acquired the statistically generated surface for WKF simulation. For the N-/P-type devices, 200 cases were generated and simulated, as shown in Figure 1d, where the regions of light color and dark color represent the low and high WKs, respectively. Figure 1e is a flow chart of the LWKF simulation. The illustration and definition of different AR devices are given in Figure 1f. The device channel has major axis "a" and minor axis "b" of different lengths of channel radius. The AR is defined as the ratio of the length of the major axis to that of the minor axis, which equals "a/b". The length of the minor axis of the ellipse-shaped channel is fixed at 5 nm, and the major axis varies with an AR of 0.5, 1, and 2, respectively, in our simulation setting. To discuss and analyze the variations that experience both WKF and variation of AR, we used a new extension of the LWKF method for the explored device with respect to different ARs [18,20] that can be implemented in device simulation. We utilized the CMOS inverter circuit consisting of N- and P-type GAA NW MOSFETs as the tested circuit to explore the timing and power fluctuations induced by WKF and the effect of AR. The schematic plot of the GAA NW CMOS inverter circuit is shown in Figure 1g. The logic input signals of the N- and P-type GAA NW MOSFETs were "1" to "0" and "0" to "1". The transition time, including rising delay time, as well as the falling delay time and the hold time of the input signal were 2, 2, and 30 ps, respectively. To estimate and capture the influence of WKF on the circuit characteristics of the explored GAA NW CMOS inverter, a coupled device–circuit simulation approach was employed, as shown in Figure 1h. This was used because a well-established equivalent circuit model of GAA NW CMOS devices is still unavailable. At first, an initial guess for device bias was assumed, and the device characteristics in the test circuit were estimated by solving the device transport equations. The obtained result was the initial guess for the coupled device–circuit simulation. Then, based on Kirchhoff's current law, the nodal equations of the tested circuits were formulated. Because the device equations were solved in the coupled device–circuit simulation, the effects of WKF on the device and the CMOS inverter circuit characteristics were thus properly captured. The coupled simulation was solved iteratively until the solution converged in each time step and bias condition.

To validate our simulation, we examined the band profile along the channel by solving 3D quantum mechanical transport and non-equilibrium Green's function models. Then, we calibrated the simulation result with measurement data of the fabricated sample [34,35]. For both the N- and P-type devices, the I_D–V_G characteristics of the simulated device at V_D = 1/−1 V were experimentally calibrated to the measured data by fitting the mobility model parameters [18,20,34,35]. Because the I_D–V_G characteristics are well-fitted between the fabrication and the simulation, this further ensures the accuracy of our statistical device and circuit simulation.

3. Results and Discussion

Figure 2 shows the standard deviation (σ) of threshold voltage, drain-induced barrier lowering (DIBL), and gate capacitance (C_G) versus AR with respect to different grain sizes of N- and P-type GAA NW MOSFETs. As the grain size reduced from 4 × 5 to 1 × 1 nm^2 and the AR induced from 0.5 to 2, σV_{th} reduced, as shown in Figure 2a,b. For a fixed channel area with a different grain size, if the grain size is large, the same gate area may contain only a few grains, so the effective WKF will be governed by high or low WKs and further lead to higher or lower V_{th}, causing relatively larger variation. Under the condition of the same grain size, the device with the larger AR has smaller fluctuation, because the grain size is relatively small. As shown in Figure 2c,d, the case of AR = 0.5 had the highest deviation, indicating that the device with the critical dimension is more sensitive to variation in the process. According to the definition of DIBL, the magnitude of σDIBL in Figure 2c,d had a similar trend to σV_{th}, as shown in Figure 2a,b, due to the dependency on V_{th}. Figure 2e,f shows the bar charts of σC_G with three different ARs and three different grain sizes. The devices with a larger AR had a larger surface area, so the value of C_G with a larger AR was larger than that of the smaller AR. However, under the condition of the same grain size, the larger AR had the smaller fluctuation. This is because the area of AR = 2 was larger, and the grain size was relatively small. Thus, the magnitude of σC_G of larger AR devices was smaller. Notably, the aspect ratio was given from a fixed axis, so it would also be helpful to interpret the result versus the device dimension using the plot of a Pelgrom model. Although we have applied the Pelgrom model to explore the variability of fin-type field-effect transistors (FinFETs) [36], the same model, we assumed, can be applied to examine the variability of a GAA NW MOSFET device.

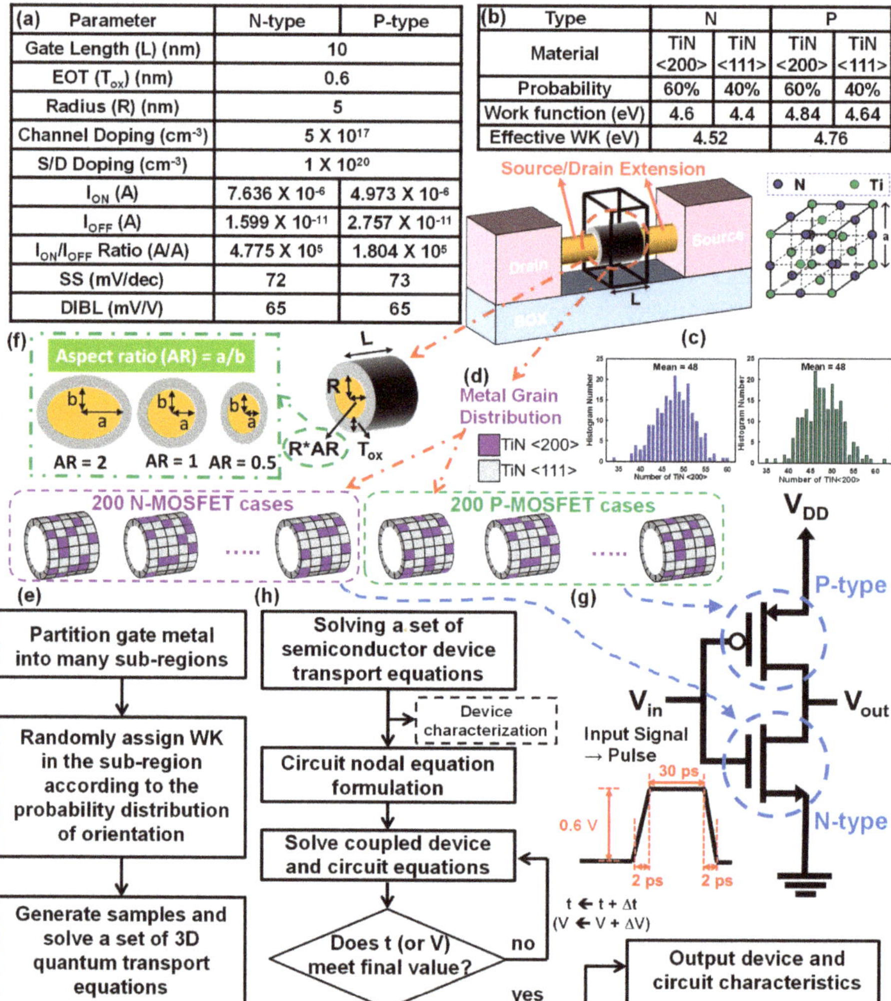

(a) Parameter	N-type	P-type
Gate Length (L) (nm)	10	
EOT (T_{ox}) (nm)	0.6	
Radius (R) (nm)	5	
Channel Doping (cm^{-3})	5×10^{17}	
S/D Doping (cm^{-3})	1×10^{20}	
I_{ON} (A)	7.636×10^{-6}	4.973×10^{-6}
I_{OFF} (A)	1.599×10^{-11}	2.757×10^{-11}
I_{ON}/I_{OFF} Ratio (A/A)	4.775×10^{5}	1.804×10^{5}
SS (mV/dec)	72	73
DIBL (mV/V)	65	65

(b) Type	N		P	
Material	TiN <200>	TiN <111>	TiN <200>	TiN <111>
Probability	60%	40%	60%	40%
Work function (eV)	4.6	4.4	4.84	4.64
Effective WK (eV)	4.52		4.76	

Figure 1. (**a**) The device's parameters and the nominal short-channel effect (SCE) values of the N-/P-type devices. We used TiN, which is a stable compound with a NaCl (sodium chloride) structure as the metal gate. According to the properties of the metal material, TiN has two different orientations: <200> and <111>, with 60% and 40% generated probabilities [19,21,22]. The related work functions (WKs) of the N-/P-type devices are shown in (**b**). (**c**) A histogram plot of the number of high WKs generated following Gaussian distribution. (**d**) The adopted test device with low and high WKs, in which the light color and dark color represent the low and high WKs, respectively. (**e**) The flow chart of work function fluctuation (WKF) simulation, where 200 cases for the N-/P-type devices were generated and simulated, as shown in (**d**). (**f**) The illustration and definition of different aspect ratio (AR) devices. (**g**) The tested complementary metal–oxide–semiconductor (CMOS) inverter circuits in this study. (**h**) Simulation flowchart for the coupled device–circuit approach. MOSFET: metal–oxide–semiconductor field-effect transistors; S: source; D: drain; DIBL: drain-induced barrier lowering; EOT: effective oxide thickness; SS: subthreshold swing; 3D: three-dimensional.

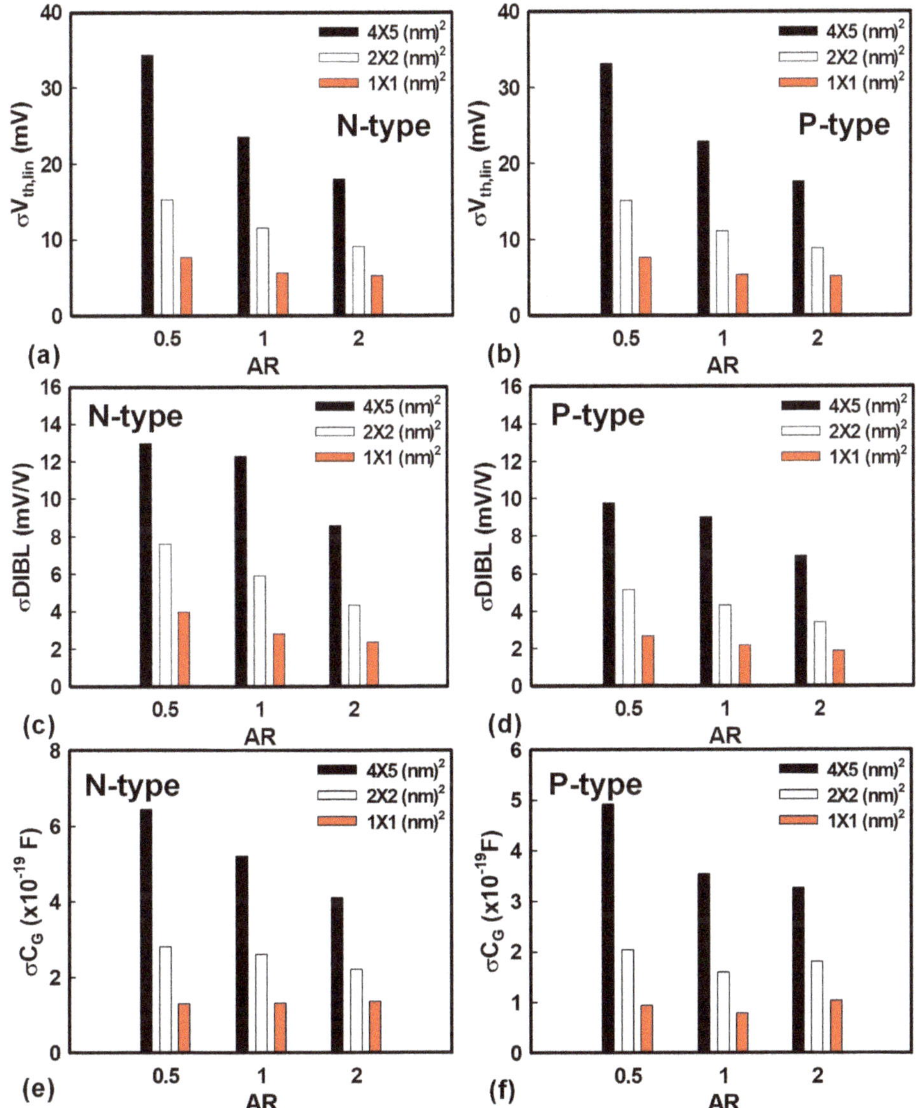

Figure 2. The standard deviation of (**a**), (**b**) V_{th}, (**c**), (**d**) DIBL, and (**e**), (**f**) gate capacitance (CG) versus different ARs with respect to different grain sizes of N-/P-type gate-all-around (GAA) nanowire (NW) MOSFETs affected by WKF. For both the N- and P-type GAA NW MOSFETs, devices with a larger grain size and smaller AR have greater deviations than the others.

Figure 3 shows the effects of a random number and random position of high WK grains on the threshold voltage: threshold voltage increases when the number of high WK grains increases. Notably, the charge distribution is strongly governed by different WKs locally. By using the LWKF method, we determined the random location effect and found that most of the high WK grains are near source (S) side or drain (D) side. Figure 3a,b shows the distributions of Case A and Case B with the highest and lowest V_{th} in the group of the same number of high WKs, respectively. The green color represents low WKs, and the white color indicates high WKs. Figure 3a′,b′ shows the corresponding conduction band energy distributions in the off-state. Because the grain pattern of Case A has a larger proportion of high WKs near the source side compared with Case B, in order to explore the difference, we illustrate the one-dimensional (1D) conduction band energy profile of the device channel center in Figure 3c. Figure 3d is a zoom-in plot and the black solid line and the red dashed line represent Case A and Case B, respectively. The barrier of Case A is 35 meV higher than that of Case B. Thus, the case with the higher barrier needs a higher voltage to lower the high barrier and make the electrons easier to pass through, leading to higher V_{th}.

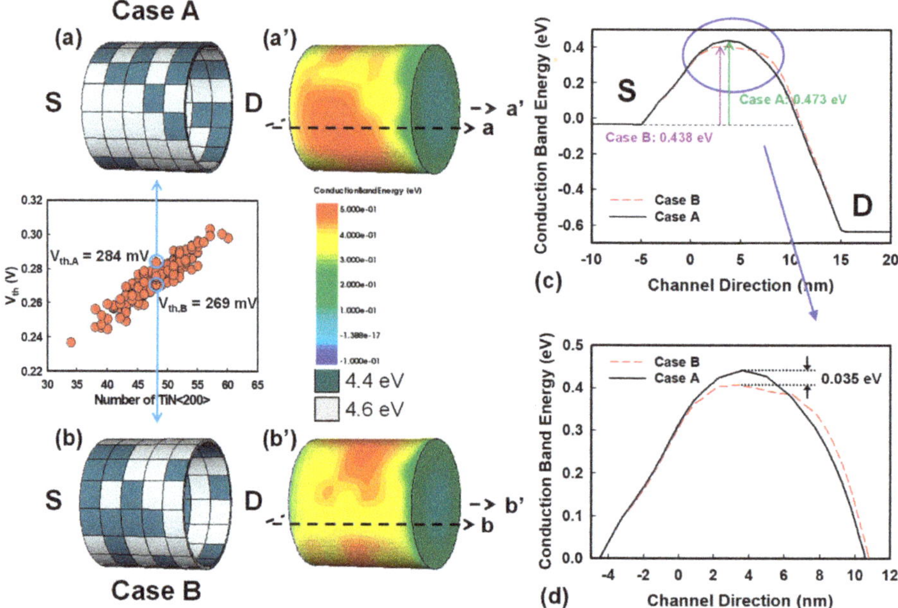

Figure 3. (**a**,**b**) The metal grain distribution of Case A and Case B that have the same number of high WKs but different V_{th}. For the metal gate, the green color represents low WKs, and the white color indicates high WKs. (**a'**,**b'**) The corresponding conduction band energy distributions in the channel region. The bias conditions are $V_D = 0.6$ V and $V_G = 0$ V. (**c**) The one-dimensional (1D) conduction band energy profile, and (**d**) the zoom-in plot from the source side to the drain side of Case A and Case B. Case A has a higher barrier in the off-state, which leads to higher V_{th} compared with Case B.

Figure 4a shows the fluctuated voltage transfer curves induced by the WKF of the explored CMOS inverter circuit. V_{IL}, the maximum permitted logic "0" at input, and V_{IH}, the minimum permitted logic "1" at input, are the extracted input voltages of the voltage transfer curves at the slope of −1V/V. These two points are used to determine NM_H and NM_L. The definition of NM is shown in Figure 4. The values of NM_H and NM_L are indicators to estimate the maximum noise signal tolerance during the operation of the inverter circuits. Figure 4b,c shows the bar chart of NM, which increases with an increasing grain size, similar to the variation of V_{th} in Figure 2a,b. Hence, NM also follows the trend of σV_{th}. Figure 4d,e displays the plots of NM_L and NM_H versus the number of high WKs affected by WKF with grain size fixed at 2×2 nm^2. When the number of high WK metals increases, NML rises and NMH does the opposite. Higher WK numbers cause a higher value of N-type V_{th} and a lower value of P-type V_{th}, resulting in both the values of V_{IL} and V_{IH} becoming higher. This leads to an increasing NM_L and a decreasing NM_H. Figure 5 shows the variance of the timing of the tested circuits experiencing WKF with three different grain sizes and three different ARs. The magnitude of variance of t_f is smaller than that of t_r owing to the larger driving capability of the N-type device. The device with the larger driving capability requires less time to charge/discharge the load capacitance. Hence, it exhibits less fall time fluctuation. The "Delay" is defined as the average of t_{HL} and t_{LH}. The larger the grain size, the larger the fluctuation of the delay time. This can be explained by the load capacitance fluctuation in Figure 5c. The σC_G of the grain equal to $4 \times 5\ nm^2$ is the largest among the three different sizes of metal grains. A larger σC_G would lead to a longer σDelay. The associated values of the timing fluctuation of different ARs are given in Figure 5d, which can verify the trend in Figure 5b—the larger the AR, the larger the timing fluctuation. Figure 6 shows t_{HL} and t_{LH} versus the number of high WKs fluctuated by WKF with grain size equal to $2 \times 2\ nm^2$. The trend of t_{HL} increases when the number of high WK metals increases, because the delay time is dependent on the start of the signal transition, which indicates the magnitude of V_{th}. Along with the rising high WK number, the value of the N-type V_{th} increases, and it becomes harder for the N-type device to turn on, causing a higher t_{HL}. For P-type devices, a larger number of high WKs leads to a lower value of V_{th}, so t_{LH} decreases. Figure 7 shows the related results of the power consumption affected by WKF and various ARs of the tested circuit.

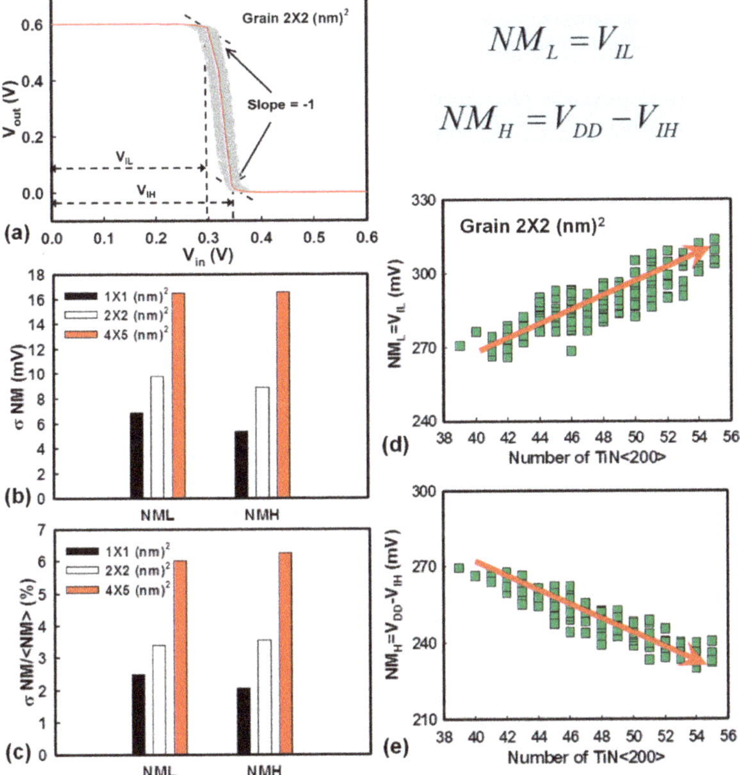

Figure 4. (**a**) The voltage transfer curves of NW MOSFET inverter circuits fluctuated by WKF. V_{IL} and V_{IH} are used to determine the noise margin (NM) of the inverter. The slope at these two points of voltage transfer curves is -1V/V. The related definitions of low level of noise margin (NML) and high level of noise margin (NMH) are shown below (**a**). (**b**,**c**) Plots of the fluctuated NM induced by the WKF of a GAA NW MOSFET with three grain sizes: $4 \times 5\ nm^2$ (red), $2 \times 2\ nm^2$ (white), and $1 \times 1\ nm^2$ (black), respectively. (**b**) The standard deviation and (**c**) the coefficient of variance of NM. (**d**,**e**) Plots of NML and NMH versus the number of TiN <200> with grain size equal to $2 \times 2\ nm^2$. The trend of NM versus high WK number between NM_L and NM_H acts conversely. When the high WK number increases, NM_L also increases, and thus, NM_H decreases.

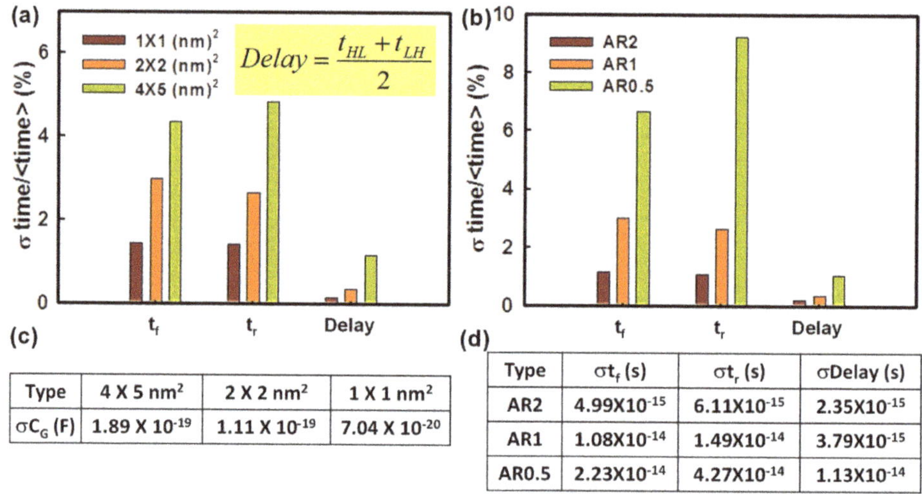

Type	4 X 5 nm²	2 X 2 nm²	1 X 1 nm²
σC_G (F)	1.89 X 10^{-19}	1.11 X 10^{-19}	7.04 X 10^{-20}

Type	σt_f (s)	σt_r (s)	σDelay (s)
AR2	4.99X10^{-15}	6.11X10^{-15}	2.35X10^{-15}
AR1	1.08X10^{-14}	1.49X10^{-14}	3.79X10^{-15}
AR0.5	2.23X10^{-14}	4.27X10^{-14}	1.13X10^{-14}

Figure 5. (**a**,**b**) The coefficient of variance of the timing analysis of GAA NW circuits experiencing WKF with different grain sizes and different ARs, respectively. (**c**) The fluctuation of load gate capacitance with three different grain sizes. (**d**) The associated values of fluctuated timing parameters of different ARs. The bias condition is $V_D = 0.05$ V and $V_G = 0.6$ V.

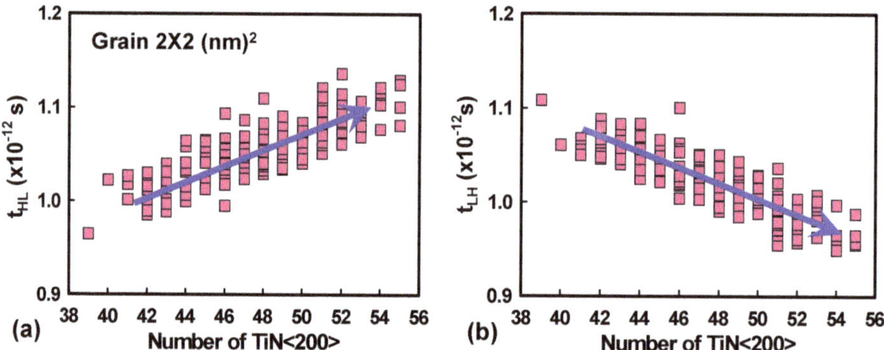

Figure 6. (**a**,**b**) The high-to-low delay time and low-to-high delay time versus the number of TiN <200> with grain size equal to 2×2 nm^2. With the increasing number of high WKs, the high-to-low delay time has become higher, and thus, the low-to-high delay time has become lower.

The total power (P_{total}) is composed of static power (P_{stat}), short-circuit power (P_{sc}), and dynamic power (P_{dyn}). The definitions of these power components are as follows:

$$P_{stat} = V_{DD} I_{leakage} \tag{1}$$

$$P_{sc} = f_{0\rightarrow 1} V_{DD} \int_T I_{sc}(\tau) d\tau \tag{2}$$

$$P_{dyn} = C_{load} V_{DD}^2 f_{0\rightarrow 1} \tag{3}$$

$$P_{total} = P_{stat} + P_{sc} + P_{dyn} \tag{4}$$

where $I_{leakage}$ is the leakage current that flows between the power rails when operating at static state. $f_{0\rightarrow 1}$ is the clock rate. I_{sc} is the short-circuit current, which is observed when both the N- and P-type devices are turned on simultaneously, resulting in a DC path between the power rails. T is the switching period. P_{stat} will consume as long as the V_{DD} is opened, regardless of the switching activity between input and output. P_{sc} is determined by I_{sc} and the time of existence of the DC path between the power rails. P_{dyn} is determined by the load capacitance (C_{load}).

Figure 7a,b shows the bar chart of power consumptions of different grain sizes. In Figure 7a, it can be observed that the average values of P_{sc} and P_{dyn} were the dominating roles in power dissipation. As shown in Figure 7b, all the power consumption terms followed the trend that the larger the grain size, the larger the fluctuation. For P_{dyn}, the device with grain size equal to 4×5 nm^2 displayed larger P_{dyn} owing to its larger C_{load} compared with the others. The device with grain size equal to 1×1 nm^2 had smaller P_{stat} than the devices with the other two grain sizes, because $I_{leakage}$ of 1×1 nm^2 was the smallest of the grain sizes. Additionally, Figure 7b shows that the magnitude of the variance of P_{stat} was the largest among all power consumption terms. However, its contribution to P_{total} was marginal. As a result, P_{total} was mainly affected by P_{sc} and P_{dyn}. Figure 7c shows the average power dissipation affected by the WKF of different ARs. The average values of P_{sc} and P_{dyn} were also much larger than that of P_{stat}. Therefore, for all AR devices, P_{sc} and P_{dyn} were the dominating factors in P_{total}. In addition, I_{sc} in the case of AR = 2 was the largest in Figure 7d, and this shows that devices with AR = 2 had the largest P_{sc}.

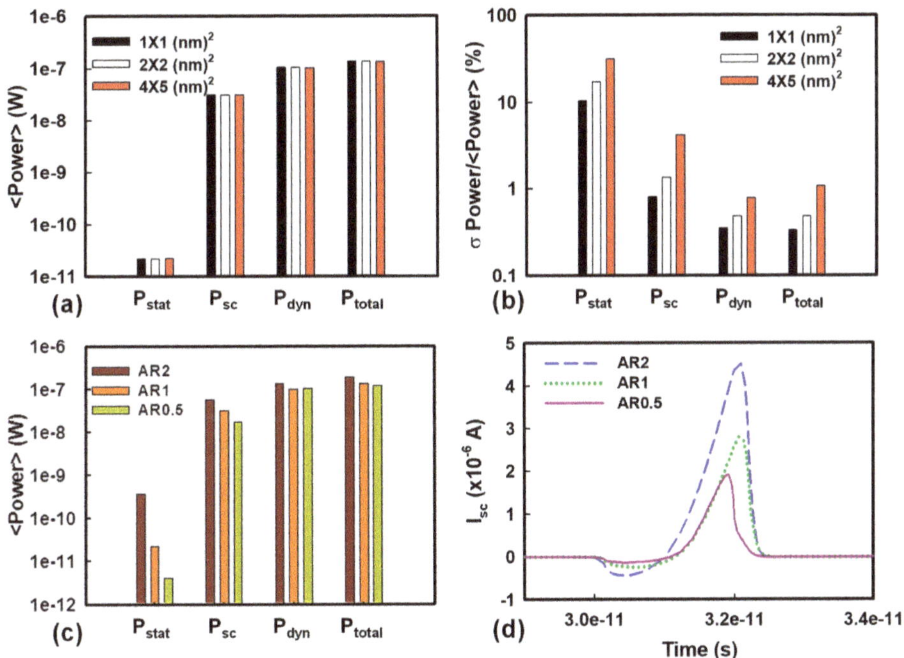

Figure 7. (**a**,**b**) Plots of each fluctuated power consumption experiencing WKF of the tested circuit with different grain sizes. (**a**) The average and (**b**) the coefficient of variance of power consumption. (**c**) The average power consumption induced by WKF of the tested circuit with different ARs. (**d**) The short-circuit current of the circuit with AR2 (blue dashed line), AR1 (green dotted line), and AR0.5 (purple solid line), respectively.

4. Conclusions

In this work, DC/AC characteristic fluctuation of GAA NW MOSFETs and variation of the dynamic property of a CMOS circuit induced by WKF and ARs of channel cross-sections were investigated using an experimentally calibrated 3D device and circuit simulation running on a parallel computing system. The V_{th} diminished with a decrease in grain size for both the N- and P-type devices. DIBL followed the trend of V_{th} due to the dependency on the V_{th} of DIBL. The standard deviation of C_G with large grain size also had greater fluctuated value. We conclude that for both DC and AC characteristics, the smaller the grain size, the lower the fluctuation. The threshold voltage increases when the number of high WK grains increases. For devices with the same number of WKs, the device with a larger proportion of high WKs near the source side will achieve to a higher threshold voltage. In addition, under the condition of same metal grain size, the larger AR device has a less severe impact from WKF than a smaller AR device, because it has a large effective gate area and the grain size is relatively small. Hence, larger AR devices will average the effect of random metal grain fluctuation and thereby reduce the degradation of WKF. For the variation of the dynamic property of the explored CMOS circuit, the delay time and NM fluctuations follow the trend of V_{th}—that a larger variation is caused by a larger grain size. For t_f and t_r, the larger driving capability of the N-type device is the reason t_f is smaller than tr. NML is positively related to the number of high WKs, while NMH is negatively related to it. In power dissipation, both P_{sc} and P_{dyn} are the most significant fluctuation sources.

Author Contributions: Conceptualization, Y.L.; Theory and Methodology, Y.L.; Programming, Software, and Simulation, Y.L., C.-Y.C., M.-H.C., and P.-J.C.; Discussion, Y.L., C.-Y.C., M.-H.C., and P.-J.C.; Draft Preparation, Y.L., C.-Y.C., M.-H.C., and P.-J.C.; Review and Editing, Y.L., C.-Y.C., and M.-H.C.

Funding: This research was funded by Ministry of Science and Technology, Taiwan grant number 106-2221-E-009-149, 106-2622-8-009-013-TM, 107-3017-F-009-001, 107-2221-E-009-094, and 107-2622-8-009-011-TM. The APC was funded by Ministry of Science and Technology, Taiwan.

Acknowledgments: This work was supported in part by the Ministry of Science and Technology (MOST), Taiwan, under grants MOST 106-2221-E-009-149, 106-2622-8-009-013-TM, 107-3017-F-009-001, 107-2221-E-009-094, and 107-2622-8-009-011-TM, and the "Center for mmWave Smart Radar Systems and Technologies" under the Featured Areas Research Center Program within the framework of the Higher Education Sprout Project by the Ministry of Education (MOE) in Taiwan.

Conflicts of Interest: The authors declare no conflicts of interest.

References

1. Li, Y.; Chang, H.-T.; Lai, C.-N.; Chao, P.-J.; Chen, C.-Y. Process Variation Effect, Metal-Gate Work-Function Fluctuation and Random Dopant Fluctuation of 10-nm Gate-All-Around Silicon Nanowire MOSFET Devices. In Proceedings of the Technical Digest of International Electron Devices Meeting, Washington, DC, USA, 7–9 December 2015.
2. Hwang, C.-H.; Li, T.-Y.; Han, M.-H.; Lee, K.-F.; Cheng, H.-W.; Li, Y. Statistical 3D Simulation of Metal Gate Workfunction Variability, Process Variation, and Random Dopant Fluctuation in Nano-CMOS Circuits. In Proceedings of the International Conference on Simulation of Semiconductor Processes and Devices, San Diego, CA, USA, 9–11 September 2009.
3. Cheng, H.-W.; Li, F.-H.; Han, M.-H.; Yiu, C.-Y.; Yu, C.-H.; Lee, K.-F.; Li, Y. 3D Device Simulation of Work Function and Interface Trap Fluctuations on High-κ/Metal Gate Devices. In Proceedings of the Technical Digest of International Electron Devices Meeting, San Francisco, CA, USA, 6–8 December 2010.
4. Tan, C.M.; Chen, X. Random dopant fluctuation in gate-all-around nanowire FET. In Proceedings of the IEEE International Nanoelectronics Conference, Sapporo, Japan, 28–31 July 2014.
5. Arnaud, F.; Pinzelli, L.; Gallon, C.; Rafik, M.; Mora, P.; Boeuf, F. Challenges and opportunity in performance, variability and reliability in sub-45 nm CMOS technologies. *Microelectron. Reliab.* **2011**, *51*, 1508–1514. [CrossRef]
6. Stathis, J.H.; Wang, M.; Southwick, R.G.; Wu, E.Y.; Linder, B.P.; Liniger, E.G.; Bonilla, G.; Kothari, H. Reliability Challenges for the 10 nm Node and Beyond. In Proceedings of the Technical Digest of International Electron Devices Meeting, San Francisco, CA, USA, 15–17 December 2014.
7. Kuhn, K.J. Considerations for Ultimate CMOS Scaling. *IEEE Trans. Electron Devices* **2012**, *59*, 1813–1828. [CrossRef]
8. Hussain, M.M.; Quevedo-Lopez, M.A.; Alshareef, H.N.; Wen, H.C.; Larison, D.; Gnade, B.; El-Bouanani, M. Thermal Annealing Effects on A Representative High-κ/Metal Film Stack. *Semicond. Sci. Technol.* **2006**, *21*, 1437–1440. [CrossRef]

9. Heu, J.L.; Setsuhara, Y.; Shimizu, I.; Miyake, S. Structure Refinement and Hardness Enhancement of Titanium Nitride Films by Addition of Copper. *Surf. Coat. Technol.* **2001**, *137*, 38–42. [CrossRef]
10. Matsukawa, T.; Liu, Y.; Endo, K.; Tsukada, J.; Yamauchi, H.; Ishikawa, Y.; O'uchi, S.; Mizubayashi, W.; Ota, H.; Migita, S.; et al. Influence of work function variation of metal gates on fluctuation of sub-threshold drain current for Fin field-effect transistors with undoped channels. *Jpn. J. Appl. Phys.* **2014**, *53*, 4S. [CrossRef]
11. Li, Y.; Cheng, H.-W.; Yiu, C.-Y.; Su, H.-W. Nanosized metal grains induced electrical characteristic fluctuation in 16-nm-gate high-κ/metal gate bulk FinFET devices. *Microelectron. Eng.* **2011**, *88*, 1240–1242. [CrossRef]
12. Kundu, A.; Koley, K.; Dutta, A.; Sarkar, C.K. Impact of gate metal work-function engineering for enhancement of subthreshold analog/RF performance of underlap dual material gate DG-FET. *Microelectron. Reliab.* **2014**, *54*, 2717–2722. [CrossRef]
13. Lee, K.-C.; Fan, M.-L.; Su, P. Investigation and comparison of analog figures-of-merit for TFET and FinFET considering work-function variation. *Microelectron. Reliab.* **2015**, *55*, 332–336. [CrossRef]
14. Seoane, N.; Indalecio, G.; Aldegunde, M.; Nagy, D.; Elmessary, M.A.; García-Loureiro, A.J.; Kalna, K. Comparison of Fin-Edge Roughness and Metal Grain Work Function Variability in InGaAs and Si FinFETs. *IEEE Trans. Electron Devices* **2016**, *63*, 1209–1216. [CrossRef]
15. Nawaz, S.M.; Mallik, A. Effects of Device Scaling on the Performance of Junctionless FinFETs Due to Gate-Metal Work Function Variability and Random Dopant Fluctuations. *IEEE Electron Device Lett.* **2016**, *37*, 958–961. [CrossRef]
16. Yang, C.-C.; Huang, W.-H.; Hsieh, T.-Y.; Wu, T.-T.; Wang, H.-H.; Shen, C.-H.; Yeh, W.-K.; Shiu, J.-H.; Chen, Y.-H.; Wu, M.-C.; et al. High Gamma Value 3D-Stackable HK/MG-Stacked Tri-Gate Nanowire Poly-Si FETs With Embedded Source/Drain and Back Gate Using Low Thermal Budget Green Nanosecond Laser Crystallization Technology. *IEEE Electron Device Lett.* **2016**, *37*, 533–536. [CrossRef]
17. Dev, S.; Meena, M.; Harsha Vardhan, P.; Lodha, S. Statistical Simulation Study of Metal Grain-Orientation-Induced MS and MIS Contact Resistivity Variability for 7-nm FinFETs. *IEEE Trans. Electron Devices* **2018**, *65*, 3104–3111. [CrossRef]
18. Li, Y.; Cheng, H.-W.; Chiu, Y.-Y.; Yiu, C.-Y.; Su, H.-W. A Unified 3D Device Simulation of Random Dopant, Interface Trap and Work Function Fluctuations on High-κ/Metal Gate Device. In Proceedings of the Technical Digest of International Electron Devices Meeting, Washington, DC, USA, 5–7 December 2011.
19. Dadgour, H.; Endo, K.; Vivek, D.; Banerjee, K. Modeling and analysis of grain-orientation effects in emerging metal-gate devices and implications for sram reliability. In Proceedings of the Technical Digest of International Electron Devices Meeting, San Francisco, CA, USA, 15–17 December 2008.
20. Li, Y.; Chen, C.-Y.; Chen, Y.-Y. Random-Work-Function-Induced Characteristic Fluctuation in 16-nm-Gate Bulk and SOI FinFETs. *Int. J. Nanotechnol.* **2014**, *11*, 1029–1038. [CrossRef]
21. Dadgour, H.F.; Endo, K.; De, V.K.; Banerjee, K. Grain-orientation induced work function variation in nanoscale metal-gate transistors—Part I: Modeling, analysis, and experimental validation. *IEEE Trans. Electron Devices* **2010**, *57*, 2504–2514. [CrossRef]
22. Indalecio, G.; García-Loureiro, A.J.; Iglesias, N.S.; Kalna, K. Study of Metal-Gate Work-Function Variation Using Voronoi Cells: Comparison of Rayleigh and Gamma Distributions. *IEEE Trans. Electron Devices* **2016**, *63*, 2625–2628. [CrossRef]
23. Li, Y.; Huang, C.-H. The Effect of the Geometry Aspect Ratio on the Silicon Ellipse-Shaped Surrounding-Gate Field-Effect Transistor and Circuit. *Semicond. Sci. Technol.* **2009**, *24*, 095018. [CrossRef]
24. Tienda-Luna, I.M.; Ruiz, F.G.; Godoy, A.; Donetti, L.; G'amiz, F. Effects of deviations in the cross-section of square nanowires. In Proceedings of the International Workshop on Computational Electronics, Pisa, Italy, 26–29 October 2010.
25. Jha, S.; Kumar, A.; Kumar, S. Impact of Elliptical Cross-Section on Some Electrical Properties of Gate-All-Around MOSFETs. *Bonfring Int. J. Power Syst. Integr. Circuits* **2012**, *2*, 18–22. [CrossRef]
26. Li, Y.; Huang, C.-H.; Li, T.-Y.; Han, M.-H. Process-Variation Effect, Metal-Gate Work-Function Fluctuation, and Random-Dopant Fluctuation in Emerging CMOS Technologies. *IEEE Trans. Electron Device* **2010**, *57*, 437–447. [CrossRef]
27. Han, M.-H.; Li, Y.; Hwang, C.-H. The impact of high-frequency characteristics induced by intrinsic parameter fluctuations in nano-MOSFET device and circuit. *Microelectron. Reliab.* **2010**, *50*, 657–661. [CrossRef]
28. Hwang, C.-H.; Li, Y.; Han, M.-H. Statistical Variability in FinFET Devices with Intrinsic Parameter Fluctuations. *Microelectron. Reliab.* **2010**, *50*, 635–638. [CrossRef]

29. Li, Y.; Hwang, C.-H.; Cheng, H.-W. Process-variation- and random-dopants-induced threshold voltage fluctuations in nanoscale planar MOSFET and bulk FinFET devices. *Microelectron. Eng.* **2009**, *86*, 277–282. [CrossRef]
30. Kuhn, K.J. CMOS Scaling for the 22 nm Node and Beyond: Device Physics and Technology. In Proceedings of the International Symposium on VLSI Technology, Systems and Applications, Hsinchu, Taiwan, 25–27 April 2011.
31. Li, Y. Optimal Geometry Aspect Ratio of Ellipse-Shaped-Surrounding-Gate Nanowire Field Effect Transistors. *J. Nanosci. Nanotechnol.* **2016**, *16*, 920–923. [CrossRef]
32. Han, K.; Hsu, P.-F.; Beach, M.; Henry, T.; Yoshida, N.; Brand, A. Metal Gate Work Function Modulation by Ion Implantation for Multiple Threshold Voltage FinFET Devices. In Proceedings of the Extended Abstracts of International Workshop on Junction Technology, Kyoto, Japan, 6–7 June 2013.
33. Han, K.; Lee, J.; Tang, S.; Maynard, H.; Yoshida, N.; Brand, A. FinFET Multi-Vt Tuning with Metal Gate Work Function Modulation by Plasma Doping. In Proceedings of the International Workshop on Junction Technology, Shanghai, China, 18–20 May 2014.
34. Yang, F.-L.; Lee, D.-H.; Chen, H.-Y.; Chang, C.-Y.; Liu, S.-D.; Huang, C.-C.; Chung, T.-X.; Chen, H.-W.; Huang, C.-C.; Liu, Y.-H.; et al. 5 nm-Gate Nanowire FinFET. In Proceedings of the Digest of Technical Papers, Symposium on VLSI Technology, Honolulu, HI, USA, 17–19 June 2004.
35. Sung, W.-L.; Li, Y. DC/AC/RF Characteristic Fluctuations Induced by Various Random Discrete Dopants of Gate-All-Around Silicon Nanowire n-MOSFETs. *IEEE Trans. Electron Devices* **2018**, *65*, 2638–2646. [CrossRef]
36. Li, Y.; Cheng, H.-W.; Hwang, C.-H. Threshold Voltage Fluctuation in 16-nm-Gate FinFETs Induced by Random Work Function of Nanosized Metal Grain. *J. Nanosci. Nanotechnol.* **2012**, *12*, 4485–4488. [CrossRef]

Article

Simulation of the Impact of Ionized Impurity Scattering on the Total Mobility in Si Nanowire Transistors

Toufik Sadi [1,*], Cristina Medina-Bailon [2], Mihail Nedjalkov [3], Jaehyun Lee [2], Oves Badami [2], Salim Berrada [2], Hamilton Carrillo-Nunez [2], Vihar Georgiev [2] and Siegfried Selberherr [3] and Asen Asenov [2]

1 Engineered Nanosystems Group, Department of Neuroscience and Biomedical Engineering, School of Science, Aalto University, P.O. Box 12200, FI-00076 Aalto, Finland
2 Device Modelling Group, School of Engineering, University of Glasgow, Glasgow G12 8LT, Scotland, UK; Cristina.MedinaBailon@glasgow.ac.uk (C.M.-B.); jaehyun1986.lee@gmail.com (J.L.); Oves.Badami@glasgow.ac.uk (O.B.); Salim.Berrada@glasgow.ac.uk (S.B.); Hamilton.Carrillo-Nunez@glasgow.ac.uk (H.C.-N.); Vihar.Georgiev@glasgow.ac.uk (V.G.); Asen.Asenov@glasgow.ac.uk (A.A.)
3 Institute for Microelectronics, TU Wien, Gusshausstrasse 27-29/E360, 1040 Wien, Austria; mixi@iue.tuwien.ac.at (M.N.); Selberherr@TUWien.ac.at (S.S.)
* Correspondence: toufik.sadi@aalto.fi

Received: 6 December 2018; Accepted: 28 December 2018; Published: 2 January 2019

Abstract: Nanowire transistors (NWTs) are being considered as possible candidates for replacing FinFETs, especially for CMOS scaling beyond the 5-nm node, due to their better electrostatic integrity. Hence, there is an urgent need to develop reliable simulation methods to provide deeper insight into NWTs' physics and operation, and unlock the devices' technological potential. One simulation approach that delivers reliable mobility values at low-field near-equilibrium conditions is the combination of the quantum confinement effects with the semi-classical Boltzmann transport equation, solved within the relaxation time approximation adopting the Kubo–Greenwood (KG) formalism, as implemented in this work. We consider the most relevant scattering mechanisms governing intraband and multi-subband transitions in NWTs, including phonon, surface roughness and ionized impurity scattering, whose rates have been calculated directly from the Fermi's Golden rule. In this paper, we couple multi-slice Poisson–Schrödinger solutions to the KG method to analyze the impact of various scattering mechanisms on the mobility of small diameter nanowire transistors. As demonstrated here, phonon and surface roughness scattering are strong mobility-limiting mechanisms in NWTs. However, scattering from ionized impurities has proved to be another important mobility-limiting mechanism, being mandatory for inclusion when simulating realistic and doped nanostructures, due to the short range Coulomb interaction with the carriers. We also illustrate the impact of the nanowire geometry, highlighting the advantage of using circular over square cross section shapes.

Keywords: nanowire field-effect transistors; silicon nanomaterials; charge transport; one-dimensional multi-subband scattering models; Kubo–Greenwood formalism; schrödinger-poisson solvers

---

1. Introduction

As the conventional CMOS technology is approaching the scaling limits, different technologies and materials are being studied extensively to extend the end of the Roadmap. One possible solution is based on different injection mechanisms such as the tunnel field effect transistor (TFET), whose most promising characteristic is reaching sub-60 mV/dec subthreshold swings (SS) [1–3]. On the other hand, the utilization of multiple gates surrounding the channel (for instance the double-gate FinFET devices in comparison to the planar fully Depleted Silicon-On-Insulator (FDSOI) devices [4,5]) improves charge control in the channel (minimizing the short-channel effects), improves the transport properties and includes the possibility of using material and strain engineering to improve the device performance. In this context, the interest in gate-all-around (GAA) nanowire transistors (NWTs) is soaring, since they are considered as a potential replacement of the FinFET CMOS technology to the ultimate scaling limits (down to 5 nm). The benefits of NWT technology are discussed extensively in literature [6–9]. In the simulation and optimisation of NWTs, the methods incorporating quantum confinement effects into semi-classical transport approaches [10–12] are becoming very attractive thanks to their lower computational demand, as compared to the purely quantum transport and atomistic models [13–15]. The continuous interest in these models is prompted by the need to accurately model silicon (Si) based NWTs, whose electronic transport properties are conditioned by multi-subband scattering at diameters smaller than 7–8 nm [12].

The importance of spatial confinement in advanced CMOS devices whose downscaling follows the Moore's law, known as More Moore transistors (e.g., NWTs or FinFETs), is significant. In such transistors, the basic properties of electronic charge transport, derived under the assumption of an ideal periodic crystal, must be revised. In this case, the Bloch theorem is no longer applicable, as charge carriers are confined typically in a cross section perpendicular to the transport direction. In such circumstances, the charge carriers (electrons or holes) cannot be considered point-like particles and their motion is constrained in the confinement plane. Also, as a result of the position-momentum uncertainty relations, the momentum of a localized carrier is not well defined. Instead, the energy is quantized into subbands and the momentum conservation is only valid in the transport direction. The rates for the corresponding multi-subband scattering processes are altered by the overlap factor of the eigenfuctions of the subbands involved in the carrier transition events [12].

Significant efforts have been invested in calculating the mobility of silicon nanowires, employing mainly the Kubo–Greenwood (KG) formalism [10,11]. Other methods have also been used to a lesser extent, such as multi-subband Monte Carlo simulators, one-dimensional (1D) multi-subband Boltzmann transport equation (BTE) solvers [12,16], and atomistic simulations [13]. Despite this progress, more work is warranted to evaluate the performance potential of NWTs with silicon channels at the scaling limit. In this work, we employ a newly developed simulation framework using the Poisson-Schrödinger solver of the Technology Computer Aided Design (TCAD) simulator GARAND [17], coupling a three-dimensional (3D) Poisson solver to two-dimensional (2D) Schrödinger solutions in the discretization planes normal to the transport direction. GARAND is in turn coupled to a standalone KG module performing multi-subband calculations of the 1D scattering rates and calculating the mobility, in order to explore the impact of the overlap factor, the different scattering mechanisms—with special attention given to ionized impurity (II) scattering—and the impact of the cross sectional shape on the electron mobility in silicon NWTs. This work focuses studying the electron transport properties in n-type devices (see e.g., [10–13,16]), as the purpose of this study is to illustrate the impact of scattering mechanisms in nanowire transistors, in general. However, our model can be readily extended to study p-type devices, and further research is warranted to explore hole transport effects [14,18] in these 1D structures. The innovation in this work is two-fold: the unique combination of simulation tools and models integrated together to accurately model confined transport properties in very small diameter NWTs, and the extensive study of physical effects impacting transport in these nanotechnology devices. As compared to other established simulators [12,13,16], our method incorporates quantum confinement effects and provides reliable low field nanowire mobility, in a transparent manner, while requiring relatively low computational cost and simulation time.

The structure of this paper is organized as follows. Section 2 gives a general overview of the mobility simulation approach, describing the simulation of the device electrostatic and multi-subband properties, and providing the details of the considered 1D (phonon, surface roughness, and ionized impurity) scattering mechanisms as well as their rate formulas. Moreover, the different equations needed in the KG formalism and the Matthiessen rule are expressed. Section 3 outlines the main results and their discussion, including a meticulous analysis of the impact of the ionized impurity scattering mechanism on the total electron mobility.

2. Simulation Method and Physics

2.1. Simulation Framework

The simulator used in this work is based on the long-channel, low-bias mobility model, whose capabilities have already been established when analyzing the impact of scattering mechanisms on nanowire performance [19]. The simulation process proceeds in three steps. In the first step, we use the coupled 3D Poisson–2D Schrödinger solver in GARAND to evaluate the electric potential and field distributions at the cross section area of a gated (long channel) NWT, and to calculate the electron densities and the details of the electronic subbands (eigenfunctions and eigenvalues) in the NWT cross section normal to the transport direction. In the second step, we utilize the potential distribution, and the corresponding eigenfunctions (and relevant subband details) to calculate the 1D transition rates for the dominant scattering mechanisms in silicon, including the modified acoustic phonon, optical phonon, ionized impurity [18] and surface roughness scattering [20]. In the final step, we use the set of multi-subband scattering mechanism rates to calculate the scattering-limited mobilities of interesting NWT structures, by adopting the KG formalism solving for the (semi-classical) BTE within the relaxation time approximation [10,11]. Before initiating the simulation process, the confinement and transport masses are extracted from first principle simulations, accounting for the impact of the cross-section and diameters of the nanowire. The effective masses are extracted from $sp^3d^5s^*$ tight-binding simulations with a Boykin parameter set [21]. Taking into account these transport effective masses, multiple cross sections of the device are simulated in GARAND. The flowchart in Figure 1 illustrates the simulation procedure. The simulation framework provides reliable mobilities at low-field near-equilibrium conditions in transistors exhibiting strong confinement effects, such as the NWTs considered here. While the main purpose of this work is to study silicon nanowires, the simulator can be easily improved to simulate nanowires based on related alloys, such as SiGe, once a model for disordered alloy scattering is implemented.

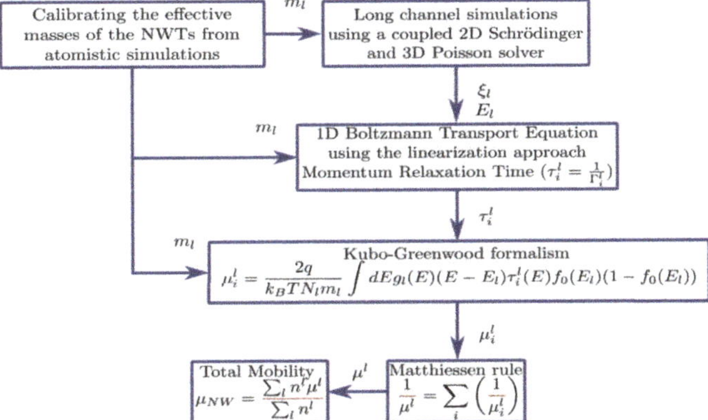

Figure 1. A flowchart illustrating the simulation framework and the corresponding steps needed to calculate the total mobility. l is the subband index; m_l, ξ_l, and E_l are the calibrated effective masses, the wavefunction, and the energy level for the lth subband, respectively; i is the i-th scattering mechanism; τ_i^l, Γ_i^l, and μ_i^l are the relaxation time, scattering rate, and mobility, respectively, for the ith mechanism and lth subband; μ_{NW} is the total mobility for a particular NW structure.

2.2. Subband Details and Electrostatics

The simulator self-consistently couples a 1D BTE model with the solution of the Poisson and Schrödinger equations in a cross section of the NWT normal to the transport direction. It involves intensive computations to determine the multi-subband energies and the eigenfunctions, and the resulting potential distribution in the NWT. This implies that the numerical aspects are an important factor for the feasibility of the simulation model, which delivers a compromise between physical accuracy and computational efficiency. GARAND couples the 2D solution of the Schrödinger equation in multiple slices (i.e., the cross section areas of the long channel) to a 3D Poisson solution in the simulated Si gate-all-around NWT, as shown in Figure 2. The solution provides the parameters and fields needed for the scattering rates and KG mobility calculations, including (i) the electric potential and field distributions used e.g., for surface roughness scattering, (ii) the details of the electronic subbands (eigenfunctions and eigenvalues) and (iii) subband electron densities.

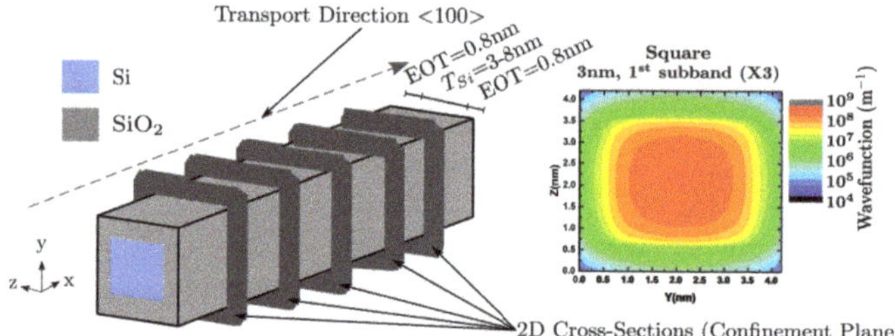

Figure 2. Coupling the 2D solution of the Schrödinger equation in multiple slices to a 3D Poisson solution in a structure based on a Si gate-all-around nanowire channel FET.

2.3. 1D Scattering Rates

The scattering rates for the electron interaction mechanisms with (acoustic and optical) phonons, impurities and surface roughness used in this study, which are considered as silicon mobility limiting mechanisms, are presented in this section. The derivation of the rates is performed within the ellipsoidal non-parabolic valleys bandstructure approximation. The rates for the scattering mechanisms are derived from the Fermi's Golden Rule, using the time-dependent perturbation theory and assuming that the transitions between two states occur instantaneously.

2.3.1. Acoustic Phonon Scattering

We consider acoustic phonon scattering mechanisms in the elastic parabolic equipartition approximation, within the short wave vector limit [22]. After extensive derivations, the scattering rate is given by:

$$\Gamma(ac,l,k) = \frac{|D_{ac}|^2 k_B T}{\rho \hbar \bar{u}^2} \frac{m}{\hbar^2} \sum_{l'} \left[\int ds |\xi_l(\mathbf{s})|^2 |\xi_{l'}(\mathbf{s})|^2 \right] \times \theta(\epsilon(k) + \Delta E_{l'}) \left(\frac{1}{|q_1 + k|} + \frac{1}{|q_2 + k|} \right), \quad (1)$$

with

$$q_{1/2} = -k \pm \sqrt{k^2 + \frac{(E_l - E_{l'})2m}{\hbar^2}}. \quad (2)$$

Here, D_{ac} is the deformation potential, k_B is the Boltzmann constant, T is the lattice temperature, ρ is the material density, $\hbar$ is the reduced Planck's constant, $\bar{u}$ is the speed of sound, m is the electron effective mass. Also, l and l' refer to the initial and final electron subbands, $\mathbf{s}$ are vectors normal to the transport direction, ξ are the wavefunctions at the given subband, θ represents the heaviside step function, $\epsilon(k)$ is the kinetic energy for a wavevctor magnitude k, and $\Delta E_{l'} = E_{l'} - E_l$ is the energy separation between subbands l and l'.

2.3.2. Optical Phonon Scattering

The energies of the different branches of the deformation potential optical phonons are approximated with constants, as used in most of the standard approaches. Accordingly, the scattering rate can be written as:

$$\Gamma(op,j,l,k)=\frac{|D_{op,j}|^2}{2\rho\omega_j}\sum_{l'}\left[\int d\mathbf{s}|\xi_l(\mathbf{s})|^2|\xi_{l'}(\mathbf{s})|^2\right]\int dqG(q), \tag{3}$$

where

$$\int dqG(q)=\frac{n_j\theta(\epsilon(k)+\Delta E^{+}_{l'j})m}{\hbar^2}\left(\frac{1}{|q_1+k|}+\frac{1}{|q_2+k|}\right)$$
$$+\frac{(n_j+1)\theta(\epsilon(k)+\Delta E^{-}_{l'j})m}{\hbar^2}\left(\frac{1}{|q_3+k|}+\frac{1}{|q_4+k|}\right)$$

with:

$$q_{1/2}=-k\pm\sqrt{k^2+\frac{(E_l+\hbar\omega_j-E_{l'})2m}{\hbar^2}},q_{3/4}=-k\pm\sqrt{k^2+\frac{(E_l-\hbar\omega_j-E_{l'})2m}{\hbar^2}}, \tag{4}$$

and

$$\Delta E^{+}_{l'j}=E_l-E_{l'}+\hbar\omega_j \quad \text{and} \quad \Delta E^{-}_{l'j}=E_l-E_{l'}-\hbar\omega_j. \tag{5}$$

Here, n_j is the equilibrium phonon number, j refers to the phonon mode and ω_j is the phonon energy.

2.3.3. Surface Roughness Scattering

Surface roughness scattering is most pronounced when confinement keeps electrons close to non-ideal interfaces. The extent of the interaction with surface imperfections is dependent on the force normal to the interface, and the statistical description of the interface roughness [20]. Assuming x is the direction of transport along the nanowire, the perturbation Hamiltonian can be written as:

$$H'=e\mathbf{E}_y(\mathbf{s},x)\Delta_y(x)+e\mathbf{E}_z(\mathbf{s},x)\Delta_z(x), \tag{6}$$

where $\Delta(y)$ and $\Delta(z)$ are the corresponding deviations of the silicon nanowire surface from the ideal surface, and $\mathbf{E}_y$ and $\mathbf{E}_z$ are the electric field components in the cross section normal to the direction of transport. The corresponding scattering rate can be derived as:

$$\Gamma(sr,l,k)=\sum_{l'}\frac{e^2}{\hbar}|N_E(l,l')|^2D^2\frac{m}{\hbar^2}\left(\frac{2\sqrt{2}\lambda}{(2+\lambda^2(k-k'_1)^2)}\frac{1}{|k'_1|}+\frac{2\sqrt{2}\lambda}{(2+\lambda^2(k-k'_2)^2)}\frac{1}{|k'_2|}\right)\theta(\epsilon_f), \tag{7}$$

where

$$N_E(l,l')=N_{\mathbf{E}_{x,y}}(l,l')=\int d\mathbf{s}\xi^*_{l'}(\mathbf{s})\mathbf{E}_{x,y}(\mathbf{s})\xi_l(\mathbf{s}). \tag{8}$$

Here, the indices x and y indicate that the integral in Equation (8) is evaluated separately for the two components of the field. The θ function comes from the requirement for existence of the square root $k'_{1,2}=\pm\sqrt{\frac{2m}{\hbar^2}(E_l+\epsilon(k)-E_{l'})}=\pm\sqrt{\frac{2m}{\hbar^2}\epsilon_f}$. The result is generalized for the particular y and z components of Equation (6). Also, D is the root mean square of the variance of Δ, and λ is the correlation length. In this work, D and λ are set to 0.48 nm and 1.3 nm, respectively.

2.3.4. Ionized Impurity Scattering

The screened Coulomb potential of an impurity at position $(\mathbf{S}, 0)$ for an electron at $(\mathbf{s}, z)$ is given by [18]:

$$V(\mathbf{r}) = \frac{Z_I e^2}{4\pi\epsilon\sqrt{(\mathbf{S}-\mathbf{s})^2 + z^2}} e^{-\left(\sqrt{(\mathbf{S}-\mathbf{s})^2+z^2}\right)/L_D}, \tag{9}$$

where, for a multi-subband case, the screening (Debye) length L_D is given by:

$$L_D^2 = \frac{kT\epsilon}{e^2 n_0} \frac{\sum_l \mathcal{F}_{-1/2}((E_l - E_F)/kT)}{\sum_l \mathcal{F}_{-3/2}((E_l - E_F)/kT)}. \tag{10}$$

Here $\mathcal{F}_n$ is the Fermi integral of order n, and n_0 is the equilibrium electron density. The scattering rate is given by

$$\Gamma(imp, \mathbf{S}', Z'; l, k) = \frac{2m}{\hbar^3} N_I \left(\frac{Z_I e^2}{4\pi\epsilon}\right)^2 \frac{1}{(2\pi)^3} \sum_{j=1}^{2} \frac{1}{2|q_j + k|} \sum_{l'} \left| \int d\mathbf{q}_S \int \frac{4\pi d\mathbf{s}\mathbf{v}\xi_{l'}^*(\mathbf{s})\xi_l(\mathbf{s})e^{i\mathbf{q}_S\mathbf{s}}}{\mathbf{q}_S^2 + q_j^2 + (1/L_D)^2} \right|^2. \tag{11}$$

In this case:

$$q_{1/2} = -k \pm \sqrt{k^2 - \frac{2m(E_{l'} - E_l)}{\hbar^2}}. \tag{12}$$

It is noteworthy that, for both surface roughness and ionized impurity scattering, we assume elastic and intra-valley transitions.

2.4. Mobility Calculation

The mobility, calculated from the set of multi-subband scattering mechanisms discussed above, is based on the KG theory applied to a confined 1D electron gas. The semi-classical simulation of the transport properties of a 1D electron gas requires the solution of the BTE, which is not straightforward in complex structures. The KG formalism employed in this study is based on the relaxation times approximation, which represents a solution of the linearized BTE. The KG formalism provides accurate mobility values at low-field near-equilibrium conditions, based on the rates of the relevant scattering mechanisms governing multi-subband transitions in quantum wires. The developed mobility theory involves a set of approximations used to define the momentum relaxation time of the 1D electron gas. For a scattering mechanism i and subband l, the mobility is calculated by

$$\mu_i^l = \frac{2q}{k_B T N_l m_l} \int dE g_l(E)(E - E_l)\tau_i^l(E) f_0(E)(1 - f_0(E)) \tag{13}$$

where m_l is the subband effective mass, τ_i^l is the relaxation time, g_l is 1D density of states, f_0 is the Fermi-Dirac function, and N_l is the '1D' electron concentration of the lth subband given by

$$N_l = N_{C1D}\left[-\mathcal{F}\{\frac{1}{2}, -\exp(\frac{E_F - E_l}{k_B T})\}\right]. \tag{14}$$

In this case, N_{C1D} is the effective 1D density of states, $\mathcal{F}$ is the Fermi integral of order 1/2, and E_F is the Fermi level. Since several scattering phenomena are considered at the same time in this study, the total mobility μ_l for each subband l must include the cumulative effect of all scattering probabilities by applying the so-called Matthiessen's rule:

$$\frac{1}{\mu^l} = \sum_i \left(\frac{1}{\mu_i^l}\right). \tag{15}$$

The average mobility of a particular nanowire structure must be calculated accounting for all the subbands:

$$\mu_{NW} = \frac{\sum_l n^l \mu^l}{\sum_l n^l}, \tag{16}$$

where n^l is the electron density in the given subband.

3. Results and Discussion

The simulated GAA structure is illustrated in Figure 2. It consists of a square (default structure) or a circular Si nanowire of a [100] channel orientation, surrounded by a thin silicon dioxide layer, with an equivalent oxide thickness (EOT) of 0.8 nm, and a gate contact. The nanowire diameter is varied from 8 nm, where the NWT has a bulk-like behaviour, down to 3 nm, where the device's behaviour is fully quantum. This work additionally presents a deeper analysis of the characteristics of a 3 nm diameter NWT, as this is the smallest possible thickness before mobility sharply falls due to scattering from confinement fluctuation.

As is known, FinFETs and NWTs have the advantage of tolerating very low doping, resulting in low statistical variability [23]. However, many NWT designs rely e.g., on the formation of junctions between the heavily doped source and drain contacts and the undoped 1D channel, imposing severe constraints on doping techniques to fulfill the need for ultrasharp source and drain junctions. In this context, understanding the impact of charged impurity scattering on the NWT behaviour is of great interest. The presented simulations are carried out for a wide range of equivalent ionized impurity/doping concentrations N_I, from very low values (10^{16} cm^{-3}) to values as high as 10^{20} cm^{-3}. The practical purpose of such analysis is to illustrate how e.g., the unintentional presence of discrete charged impurities [24] or intentional doping during the fabrication process of Si NW channels [25,26] (for example for threshold voltage tuning), can affect the device performance. Indeed, scattering by ionized impurity (Coulombic) centers is suggested to be very important in very thin Si NWs with diameters below 5 nm, due to their small cross section [18,27].

Figure 3a shows the variation of the rates for acoustic and optical phonons, as well as surface roughness scattering as a function of the total energy, for the above NWT with a diameter of 3 nm and at a moderately high sheet density of 3.25×10^{12} cm^{-2}. In this paper, we assume that the sheet density N_s in the nanowire (of any shape) is related to the volumetric density N via $N = N_s/L_p$, where L_p is the nanowire perimeter. Figure 3b shows the corresponding variation of the ionized impurity rates at channel doping concentrations varying from $\sim 2 \times 10^{17}$ cm^{-3} to $\sim 10^{19}$ cm^{-3}, and the corresponding free carrier sheet densities assuming charge neutrality. The carrier sheet density is controlled directly by varying the bias of the gate contact V_G, increasing as V_G is raised. The rates are shown for all the subbands in the three two-fold conduction valleys of interest in the [100] oriented NWT channel. While the impact of surface roughness scattering at low carrier sheet densities is small, its rate becomes significant at the charge density of 3.25×10^{12} cm^{-2} as highlighted in Figure 3a. While the acoustic phonon scattering rate is in general higher than the optical phonon scattering rate, the impact of the former mechanism on mobility is expected to be much more pronounced, as its rate is many times higher at the lowest subband where most electrons are located. As for ionized impurity, its rate increases at higher impurity concentrations, as expected. For the impurity concentrations mentioned above ($>\sim 10^{17}$ cm^{-3}), the II rate is generally highest compared to all other scattering mechanisms at the lowest subband where most electrons are located, making II scattering one of the dominant factors impacting negatively NWT mobility, as discussed in more detail below.

Figure 4a shows the ionized impurity limited electron mobility as a function of the sheet density considering various ionized impurity densities (N_I) as well as charge neutrality ($N_I = n_0$). As the II scattering rate is directly proportional to the impurity density, the II limited mobility decreases according to $1/N_I$. This is reflected in the curve showing the rate variation for the charge neutrality case where the mobility curve has a hyperbola-like shape when N_I is varied. The II-limited mobility falls from 4000 cm^2/Vs at $N_I = 10^{16}$ cm^{-3} to below 60 cm^2/Vs at $N_I = 10^{20}$ cm^{-3}, indicating that this scattering mechanism is dominant at high N_I, adversely affecting the mobility of the nanowire structure. Figure 4b shows the electron mobility as a function of the sheet density considering acoustic (Ac Ph), optical (Op Ph), total phonon (Ph), surface roughness (SR), and ionized impurity (assuming charge neutrality) scattering separately, as well as the combined mechanisms. The phonon limited mobility is almost constant up to a sheet density of $\sim 3 \times 10^{12}$ cm^{-2}, and starts to vary and visibly decrease at higher densities. The surface roughness limited mobility is relatively constant up to a sheet density of $\sim 3 \times 10^{11}$ cm^{-2}. However, it starts to decrease dramatically as the sheet density is increased reaching a value much lower than the phonon-limited mobility at $\sim 10^{13}$ cm^{-2}; this significant decrease is partly associated with the higher magnitudes of the field components perpendicular to the transport direction. As expected, II scattering may degrade the total mobility especially at sheet densities beyond $\sim 10^{11}$ cm^{-2}. At the highest sheet density considered, the phonon-limited mobility is 200 cm^2/Vs; with surface roughness and II scattering included, the mobility falls down to 50 cm^2/Vs and 20 cm^2/Vs, respectively. It is noteworthy that, at very high sheet densities (>4 $\times$ 10^{12} cm^{-2}) and impurity concentrations (>10^{20} cm^{-3}), the II scattering starts to decrease again, due to screening, explaining the visible increase in the II-limited mobility shown at these densities in Figure 4.

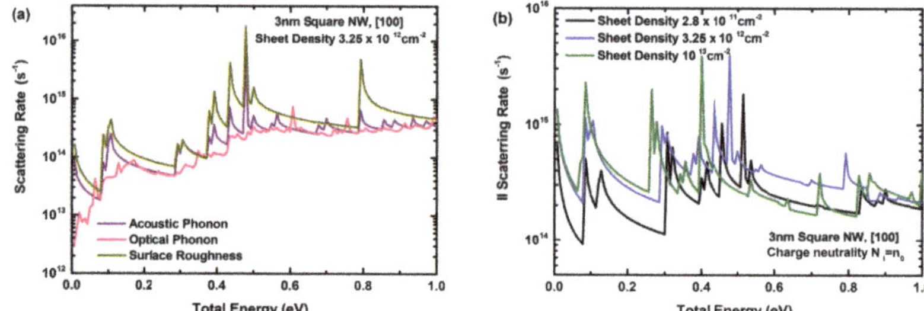

Figure 3. (**a**) The rates for acoustic and optical phonons, as well as surface roughness scattering as a function of the total energy, for a sheet density of 3.25 $\times$ 10^{12} cm^{-2}. (**b**) The rates for ionized impurity scattering, as a function of the total energy and different sheet densities, assuming charge neutrality. The results are for a 3 nm diameter square nanowire structure.

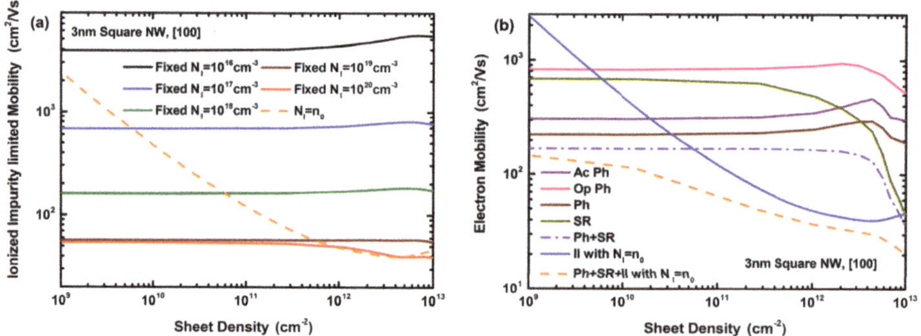

Figure 4. (**a**) Ionized impurity limited electron mobility as a function of the sheet density considering different fixed ionized impurity densities as well as charge neutrality. (**b**) Electron mobility as a function of the sheet density considering acoustic, optical, total phonon, surface roughness, and ionized impurity (assuming charge neutrality) scattering separately, as well as the combined mechanisms. The results are for a 3 nm diameter square nanowire structure.

Figure 5 shows how the total electron mobility varies, as a function of sheet density, for different ionized impurity densities and the charge neutrality case. Figure 5 confirms the role of II scattering in degrading the total NWT channel mobility for a large range of sheet densities. Figure 6 illustrates the variation of the nanowire mobility with the nanowire cross sectional area, for three cases: when including (a) only phonon and SR (Ph + SR) scattering, and phonon and SR scattering alongside ionized impurity scattering with (b) a fixed (moderately high) impurity concentration ($N_I = 10^{18}$ cm^{-3}) and (c) the charge neutrality case. In all cases, circular nanowires exhibit generally higher mobilities than the square nanowires for the same area. Albeit, such an advantage is less significant when including II scattering, especially at higher impurity concentrations as is the case for lower diameters (e.g., 3 nm and 4 nm) in the charge neutrality case.

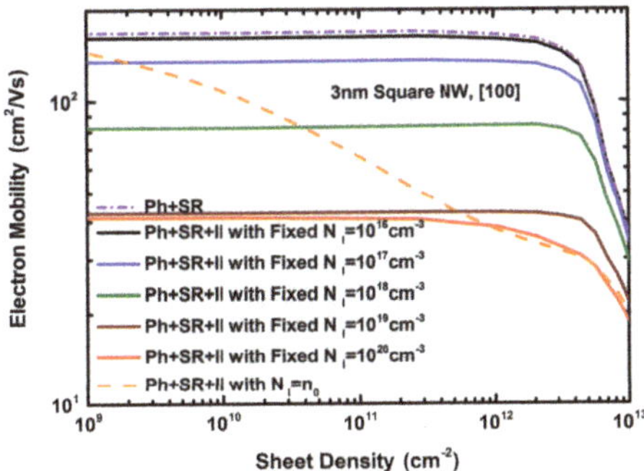

Figure 5. The electron mobility as a function of the sheet density considering the combined total phonon and surface roughness (Ph + SR) scattering rates, as well as the total scattering (Ph + SR + II). The results are for different fixed ionized impurity densities and the charge neutrality case, for a 3 nm diameter square NW with a [100] channel orientation.

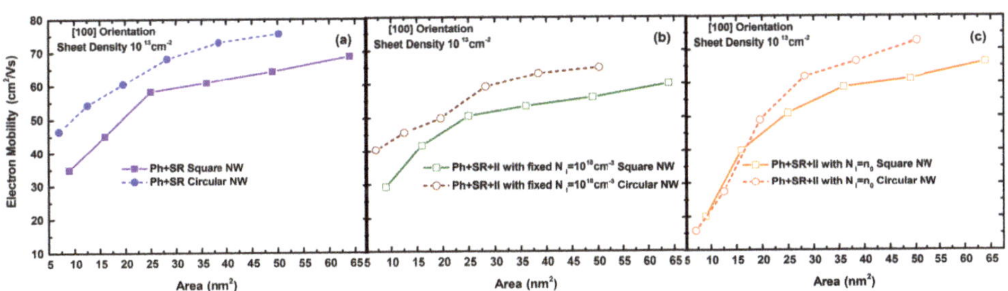

Figure 6. The electron mobility as a function of the NW cross sectional area, considering the impact of the combined total phonon and surface roughness (Ph + SR) scattering effects (**a**), as well as all present scattering mechanisms (Ph + SR + II) with a (**b**) fixed ionized impurity density ($N_I = 10^{18}$ cm^{-3}) and (**c**) the charge neutrality case. The results are for both square and circular NWs ([100] orientation) and a sheet density of 10^{13} cm^{-2}.

4. Conclusions

We have developed and employed a unique multi-subband simulator, coupling the KG formalism to an efficient Poisson–Schrödinger solver, to provide a deeper insight into charge transport in NWTs. We have explored the effect of various scattering mechanisms and nanowire shapes on the electron mobility of gate-all-around Si NWTs. This paper emphasizes the importance of accounting correctly for quantum confinement and multi-subband transistions, to provide a reliable prediction of device performance, and the role of phonon, surface roughness and ionized impurity scattering in degrading the nanowire mobility. Results show that acoustic phonon scattering is the dominant phonon scattering mechanism, and that the influence of surface roughness scattering becomes significant compared to phonon-mediated mechanisms only at very high mobile charge densities. Also, we carried out a thorough analysis of the (negative) impact of the ionized impurity scattering, showing that this mechanism dominates in realistic and doped structures, at impurity concentrations above $\sim 10^{18}$ cm^{-3}. Further analysis shows that circular nanowires give higher mobilities compared to square nanowires, at a given cross-sectional area, but such benefit is minimized at high impurity concentrations when including II scattering. The presented simulation framework is a valuable tool for technology option screening for ultra-scaled NWTs, in terms of performance.

Author Contributions: Research conceptualization, T.S., C.M.-B., M.N., S.S., and A.A.; Data curation, C.M.-B., J.L., O.B., S.B. and H.C.-N.; Formal analysis, T.S., C.M.-B. and M.N.; Funding acquisition, M.N., S.S. and A.A.; Investigation, T.S., C.M.-B., M.N., V.G., S.S. and A.A.; Methodology, T.S., C.M.-B., M.N., J.L., O.B., S.B., H.C.-N. and V.G.; Project administration, T.S., C.M.-B., M.N., S.B., V.G., S.S. and A.A.; Software, T.S., C.M.-B., J.L., O.B., S.B., H.C.-N. and V.G.; Supervision, M.N., S.S. and A.A.; Validation, T.S., C.M.-B., M.N., S.S. and A.A.; Writing—original draft, T.S.; Writing—review & editing, T.S., C.M.-B., M.N. S.S. and A.A.

Funding: This work was supported by the European Union's Horizon 2020 research and innovation programme under grant agreement No 688101, within the SUPERAID7 project.

Acknowledgments: The authors have used various computational facilities available at the University of Glasgow, TU Wien and Aalto University. Collaboration between these institutions was made possible thanks to the aforementioned EU SUPERAID7 project and its network.

Conflicts of Interest: The authors declare no conflict of interest.

References

1. Ionescu, A.M.; Riel, H. Tunnel field-effect transistors as energy-efficient electronic switches. *Nature* **2011**, *479*, 329–337. [CrossRef] [PubMed]
2. Björk, M.T.; Knoch, J.; Schmid, H.; Riel, H.; Riess, W. Silicon nanowire tunneling field-effect transistors. *Appl. Phys. Lett.* **2008**, *92*, 193504. [CrossRef]
3. Imenabadi, R.M.; Saremi, M.; Vandenberghe, W.G. A Novel PNPN-Like Z-Shaped Tunnel Field- Effect Transistor With Improved Ambipolar Behavior and RF Performance. *IEEE Trans. Electron Devices* **2017**, *64*, 4752–4758. [CrossRef]
4. Saremi, M.; Afzali-Kusha, A.; Mohammadi, S. Ground plane fin-shaped field effect transistor (GP-FinFET): A FinFET for low leakage power circuits. *Microelectron. Eng.* **2012**, *95*, 74–82. [CrossRef]
5. Khandelwal, S.; Duarte, J.P.; Khan, A.I.; Salahuddin, S.; Hu, C. Impact of parasitic capacitance and ferroelectric parameters on negative capacitance FinFET characteristics. *IEEE Electron Device Lett.* **2017**, *38*, 142–144. [CrossRef]
6. Ferry, D.K.; Gilbert, M.J.; Akis, R. Some considerations on nanowires in nanoelectronics. *IEEE Trans. Electron Devices* **2008**, *55*, 2820–2826. [CrossRef]
7. Appenzeller, J.; Knoch, J.; Bjork, M.T.; Riel, H.; Schmid, H.; Riess, W. Toward nanowire electronics. *IEEE Trans. Electron Device* **2008**, *55*, 2827–2845. [CrossRef]
8. Yu, B.; Wang, L.; Yuan, Y.; Asbeck, P.M.; Taur, Y. Scaling of nanowire transistors. *IEEE Trans. Electron Device* **2008**, *55*, 2846–2858. [CrossRef]
9. Lu, W.; Xie, P.; Lieber, C.M. Nanowire transistor performance limits and applications. *IEEE Trans. Electron Device* **2008**, *55*, 2859–2876. [CrossRef]
10. Tienda-Luna, I.M.; Ruiz, F.G.; Godoy, A.; Biel, B.; Gamiz, F. Surface roughness scattering model for arbitrarily oriented silicon nanowires. *J. Appl. Phys.* **2011**, *110*, 084514. [CrossRef]
11. Jin, S.; Fischetti, M.V.; Tang, T.-W. Modeling of electron mobility in gated silicon nanowires at room temperature: Surface roughness scattering, dielectric screening, and band nonparabolicity. *J. Appl. Phys.* **2007**, *102*, 083715. [CrossRef]
12. Sadi, T.; Towie, E.; Nedjalkov, M.; Riddet, C.; Alexander, C.; Wang, L.; Georgiev, V.; Brown, A.; Millar, C.; Asenov, A. One-dimensional multi-subband Monte Carlo simulation of charge transport in Si nanowire transistors. In Proceedings of the Simulation of Semiconductor Processes and Devices 2016, Nuremberg, Germany, 5 October 2016; pp. 23–26. [CrossRef]
13. Zhang, W.; Delerue, C.; Niquet, Y.-M.; Allan, G.; Wang, E. Atomistic modeling of electron-phonon coupling and transport properties in n-type [110] silicon nanowires. *Phys. Rev. B* **2010**, *82*, 115319. [CrossRef]

14. Niquet, Y.-M.; Delerue, C.; Rideau, D.; Videau, B. Fully atomistic simulations of phonon-limited mobility of electrons and holes in ⟨001⟩-, ⟨110⟩-, and ⟨111⟩-oriented Si nanowires. *IEEE Trans. Electron Device* **2012**, *59*, 1480–1487. [CrossRef]
15. Niquet, Y.-M.; Delerue, C.; Krzeminski, C. Effects of strain on the carrier mobility in silicon nanowires. *Nano Lett.* **2012**, *12*, 3545–3550. [CrossRef] [PubMed]
16. Ramayya, E.B.; Vasileska, D.; Goodnick, S.M.; Knezevic, I. Electron transport in silicon nanowires: The role of acoustic phonon confinement and surface roughness scattering. *J. Appl. Phys.* **2008**, *104*, 063711. [CrossRef]
17. GARAND User Guide. Synopsys, Inc. 2017. Available online: https://solvnet.synopsys.com (accessed on 31 December 2018).
18. Neophytou, N.; Kosina, H. Atomistic simulations of low-field mobility in Si nanowires: Influence of confinement and orientation. *Phys. Rev. B* **2011**, *84*, 085313–085328. [CrossRef]
19. Medina-Bailon, C.; Sadi, T.; Nedjalkov, M.; Lee, J.; Berrada, S.; Carrillo-Nunez, H.; Georgiev, V.; Selberherr, S.; Asenov, A. Study of the 1D scattering mechanisms' impact on the mobility in Si nanowire transistors. In Proceedings of the 2018 Joint International EUROSOI Workshop and International Conference on Ultimate Integration on Silicon (EUROSOI-ULIS), Granada, Spain, 19–21 March 2018; doi: 10.1109/ULIS.2018.8354723.
20. Nedjalkov, M.; Ellinghaus, P.; Weinbub, J.; Sadi, T.; Asenov, A.; Dimov, I.; Selberherr, S. Stochastic analysis of surface roughness models in quantum wires. *Comput. Phys. Commun.* **2018**, *228*, 30–37. [CrossRef]
21. Boykin, T.B.; Klimeck, G.; Oyafuso, F. Valence band effective-mass expressions in the $sp^3d^5s^*$ empirical tight-binding model applied to a Si and Ge parametrization. *Phys. Rev. B* **2004**, *69*, 115201. [CrossRef]
22. Jacoboni, C.; Reggiani, L. The Monte Carlo method for the solution of charge transport in semiconductors with applications to covalent materials. *Rev. Mod. Phys.* **1983**, *55*, 645–705. [CrossRef]
23. Asenov, A. Random dopant induced threshold voltage lowering and fluctuations in sub-0.1 μm MOSFETs: A 3-D atomistic simulation study. *IEEE Trans. Electron Device* **1998**, *45*, 2505–2513. [CrossRef]
24. Akhavan, N.D.; Jolley, G.; Umana-Membreno, G.A.; Antoszewski, J.; Faraone, L. Discrete Dopant Impurity Scattering in p-Channel Silicon Nanowire Transistors: A k.p Approach. *IEEE Trans. Electron Device* **2014**, *61*, 386–393. [CrossRef]
25. Vallett, A.L.; Minassian, S.; Kaszuba, P.; Datta, S.; Redwing, J.M.; Mayer, T.S. Fabrication and Characterization of Axially Doped Silicon Nanowire Tunnel Field-Effect Transistors. *Nano Lett.* **2010**, *10*, 4813–4818. [CrossRef] [PubMed]
26. Ma, D.D.D.; Lee, C.S.; Au, F.C.K.; Tong, S.Y.; Lee, S.T. Small-Diameter Silicon Nanowire Surfaces. *Science* **2003**, *299*, 1874–1877. [CrossRef] [PubMed]
27. Markussen, T.; Rurali, R.; Jauho, A.-P.; Brandbyge, M. Scaling Theory Put into Practice: First-Principles Modeling of Transport in Doped Silicon Nanowires. *Phys. Rev. Lett.* **2007**, *99*, 076803. [CrossRef] [PubMed]

Article

InGaAs FinFETs Directly Integrated on Silicon by Selective Growth in Oxide Cavities

Clarissa Convertino *, Cezar Zota, Heinz Schmid *, Daniele Caimi, Marilyne Sousa, Kirsten Moselund and Lukas Czornomaz

IBM Research GmbH Zürich, Säumerstrasse 4, CH-8803 Rüschlikon, Switzerland; zot@zurich.ibm.com (C.Z.); cai@zurich.ibm.com (D.C.); sou@zurich.ibm.com (M.S.); kmo@zurich.ibm.com (K.M.); luk@zurich.ibm.com (L.C.)

* Correspondence: ino@zurich.ibm.com (C.C.); sih@zurich.ibm.com (H.S.)

Received: 30 November 2018; Accepted: 22 December 2018; Published: 27 December 2018

Abstract: III-V semiconductors are being considered as promising candidates to replace silicon channel for low-power logic and RF applications in advanced technology nodes. InGaAs is particularly suitable as the channel material in n-type metal-oxide-semiconductor field-effect transistors (MOSFETs), due to its high electron mobility. In the present work, we report on InGaAs FinFETs monolithically integrated on silicon substrates. The InGaAs channels are created by metal–organic chemical vapor deposition (MOCVD) epitaxial growth within oxide cavities, a technique referred to as template-assisted selective epitaxy (TASE), which allows for the local integration of different III-V semiconductors on silicon. FinFETs with a gate length down to 20nm are fabricated based on a CMOS-compatible replacement-metal-gate process flow. This includes self-aligned source-drain n^+ InGaAs regrown contacts as well as 4 nm source-drain spacers for gate-contacts isolation. The InGaAs material was examined by scanning transmission electron microscopy (STEM) and the epitaxial structures showed good crystal quality. Furthermore, we demonstrate a controlled InGaAs digital etching process to create doped extensions underneath the source-drain spacer regions. We report a device with gate length of 90 nm and fin width of 40 nm showing on-current of 100 μA/μm and subthreshold slope of about 85 mV/dec.

Keywords: III-V; TASE; MOSFETs; Integration

---

1. Introduction

Compound semiconductors based on arsenides ($In_{1-x}Ga_xAs$) [1] are considered promising candidates to replace silicon in nFETs for advanced and ultra-scaled CMOS technology nodes. These materials offer a significant advantage in terms of electron mobility compared to silicon and are suitable for low-power applications [2–5]. Nevertheless, to enable large-scale and cost-effective integration, the challenge of transferring high quality III-V material on silicon must be overcome. Recently, different strategies for III-V on silicon integration have been proposed. Strain-relaxed buffer layer growth and direct wafer bonding (DWB) [6,7], can enable large-area III-V-on-insulator substrates, as well as 3D heterogenous integration on processed substrates [8]. Selective epitaxial techniques make possible, instead, local integration of III-V crystals in pre-defined regions [9,10]. This approach can potentially reduce the costs associated with III-V substrates and simplify the integration process. Defect density can be engineered by tuning multiple aspects such as cavity geometry, nucleation seed or growth direction. Aspect-ratio trapping (ART) technique [11,12], for instance, aims to filter crystals defects propagating along (111) planes but lacking in defects confinement along the trench direction. We have previously developed an integration approach called template-assisted selective epitaxy (TASE) [13–15], based on the growth of different III-V materials within arbitrarily shaped oxide cavities. This technique has been employed to demonstrate various devices such as tunnel FETs [16,17] ballistic nanowires [18] as well as optically active devices [19].

In this work we demonstrate InGaAs n-FinFETs integrated on a silicon (100) substrate. Due to the presence of a buried oxide layer (BOX), our InGaAs-on-insulator devices share the benefits as SOI technologies. Furthermore, we implement here a replacement-metal-gate process (RMG) on selectively grown structures with a CMOS compatible III-V process. The fabrication flow includes n-doped InGaAs contacts regrowth as well as SiN_x spacers. Source-drain doped extensions are obtained by digital etching of the InGaAs channel and regrowth, in order to mitigate the access resistance increase introduced by the presence of spacers. InGaAs material quality is investigated through STEM analysis (JEOL ARM-200F, Tokyo, Japan). Electrical characteristic of FinFETs devices at two different gate length is shown, as well as the effect of scaling on the transistor subthreshold operation.

2. Materials and Methods

The InGaAs devices are fabricated by TASE. First, a SiO_2 thermal oxide layer 50 nm thick is deposited on a silicon wafer (Figure 1a). Seed-area openings, with a diameter of 50 nm, are patterned on the oxide layer by e-beam lithography using PMMA resist (Figure 1b). A sacrificial layer 30 nm thick is then deposited on top of the BOX. Thickness and morphology of this layer will define the features of the final grown semiconductor structure. Next, the sacrificial material is patterned by e-beam lithography and the structures to be transferred into III-V are dry etched (Figure 1c). A second oxide layer, 100 nm thick, referred to as the oxide template, is deposited above the patterned structures. Afterward, openings are patterned by e-beam lithography using PMMA resist and vias are dry etched down to the sacrificial material layer (Figure 1d). The latter is therefore selectively etched, exposing the silicon seed that was previously formed (Figure 1e). Thus, the as-formed cavities contain a small seed opening to the silicon substrate. The nucleation point is perpendicular to the growth direction (Figure 1f), enabling efficient defect filtering in multiple directions [15]. A short dip in diluted HF is performed to remove the native oxide formed on the Si seed. The sample is then immediately loaded into a metal–organic-chemical-vapor-deposition (MOCVD) reactor and $In_{0.53}Ga_{0.47}As$ is grown into the patterned structures. Trimethylgallium (TMGa), trimethylindium (TMIn) and tertiarybutylarsine (TBAs) are used as precursors for the InGaAs growth, performed at 550 °C. The growth is geometrically confined into the formed template cavities. Afterwards, the template is removed by a combination of dry reactive ion etching (RIE) and HF wet etching. The replacement-metal-gate (RMG) process starts with the deposition and patterning of a dummy gate (Figure 2a). 3 nm Al_2O_3 (working as etch-stop layer) and 150 nm amorphous silicon are deposited and patterned by using HSQ resist. The smallest physical gate length measured is 20 nm. The amorphous silicon is dry etched by inductively-coupled-plasma (ICP) RIE with an optimized process for vertical sidewalls. The silicon etching stops on the Al_2O_3 layer. Next, 4 nm thick SiN_x spacers are deposited and dry etched by RIE. The Al_2O_3 on the InGaAs areas not covered by the dummy gate is then removed by wet etching in HF solution. As schematized in Figure 2c, the InGaAs channel is recessed underneath the sidewall spacers. This is achieved by several cycles of III-V digital etching (DE). The cycle consists in placing the sample in ozone atmosphere for 8 min and subsequently removing the as-formed III-V oxide in an HCl dilution with de-ionized water (1:10) for 15 s. The estimated etch rate per DE cycle is about 1.5 nm. Following, the sample is immediately loaded in the MOCVD reactor to perform raised-source-drain (RSD) epitaxy of n-doped $In_{0.53}Ga_{0.47}As$. The dopant used is Sn and the estimated doping level is 1×10^{19} cm^{-3}. A SEM (Hitachi SU8000, Tokyo, Japan) micrograph of the InGaAs device after RSD epitaxy is shown in Figure 2d. RSD epitaxy is followed by an inter-layer dielectric deposition (250 nm of SiO_2) that is subsequently planarized by chemical mechanical polishing (CMP). The oxide planarization allows to expose the top part of the dummy gate that is then removed by a selective XeF_2-based etch. The dummy high-k oxide is hence removed by HF etching and the sample is immediately loaded in the ALD chamber for the gate-stack deposition. The gate-stack consists of 5 nm Al_2O_3/HfO_2 bilayer high-k dielectric and TiN metal-gate. Right after, 150 nm of W are sputtered and planarized by CMP. A second ILD (Interlayer dielectric) layer (50 nm of SiO_2) is deposited and vias down to source/drain/metal are opened and filled with tungsten. Metal pads are patterned with negative e-beam resist and dry etched with RIE. Prior to measuring the devices, a forming gas anneal (FGA) at 300 °C is performed.

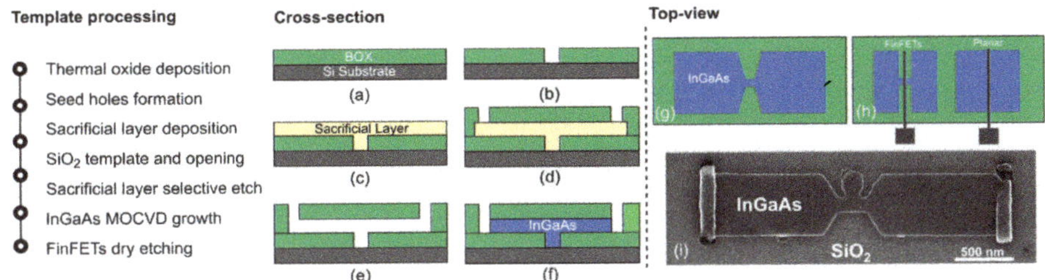

Figure 1. Overview of the fabrication process. (left) process flow up to the InGaAs fin formation. Cross-section schematic of (**a**) thermal oxide formation on silicon substrate, (**b**) patterning and opening of seed area in the oxide layer, (**c**) sacrificial layer deposition, (**d**) oxide template deposition and patterning of opening areas, (**e**) selective removal of sacrificial material, (**f**) III-V MOCVD growth. Top-view schematic of (**g**) as-grown InGaAs structure after SiO_2 template strip and (**h**) planar and FinFETs structures after dry etching. (**i**) SEM top-view image of grown InGaAs structure from a seed positioned off-center.

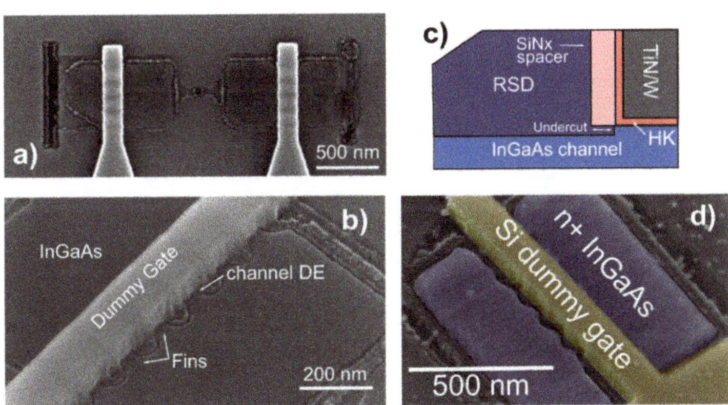

Figure 2. Device fabrication steps. (**a**) Top-view SEM image of InGaAs FinFET device after dummy gate patterning and dry etch. Fins are visible underneath the gate. (**b**) SEM picture showing InGaAs fins after digital etching (DE). The channel recess underneath the dummy gate is performed to allow for doped extensions regrowth. This is schematized in (**c**) with a cross-section drawing zooming on the channel/RSD interface. (**d**) InGaAs transistor SEM top-view after MOCVD RSD growth.

3. Results and Discussion

A schematic of the device cross-section is shown in Figure 3a with the corresponding STEM cross-sections of the finished device shown in Figure 3b and a magnified view of the spacer/RSD/channel corner (Figure 3c). The RSD and channel interface are clearly distinguishable. This is possibly due to an undesired drift in indium content between the two growth runs. The presence of planar stacking faults along a (111) plane are visible as well, originating from the growth process. Extensive study of growth dynamic and defects in InGaAs structures grown by TASE can be found in [20]. Electrical I_{DS}-V_G characteristics of 90 nm and 150 nm gate length devices as measured after FGA are shown in Figure 4. The shorter gate length (L_G) device with fin width of W_{FIN} = 40 nm, exhibits 85 mV/dec and 95 mV/dec subthreshold slopes (SS) at V_{DS} = 0.05 V and V_{DS} = 0.5 V, respectively.

SS is overall improved compared to devices reported in [15], where a SS of 190 mV/dec is achieved for L_G = 100 nm and W_{FIN} = 50 nm. The same device shows a transconductance peak value in saturation of 350 µS/µm and a maximum transconductance efficiency around 20 V^{-1} at 0.3 V. The R_{ON} of this device, measured from the output characteristics, is about 1 kΩ·µm. Compared to our previously reported InGaAs FinFETs [15] fabricated using a similar scheme, we achieved significant improvement in off-state performance as well as better $I_{ON}/I_{OFF} = 10^4$. This can be attributed to the presence of extended source-drain RSD contacts reaching underneath the spacer region, limiting the access resistance increase introduced by ungated channel area. The short L_G device shows an I_{ON} = 100 µA/µm, at fixed I_{OFF} = 100 nA/µm at V_{DS} = 0.5 V (Figure 4a) and a V_{GS} voltage swing of 0.5 V, which is the intended bias point for III-V MOSFETs. The two reported L_G, with a comparable slope, illustrate the positive impact of scaling on the on-state performance. The transfer characteristic of a planar device is shown in Figure 4b. For the same L_G, planar devices show higher SS due to degraded electrostatic control, and lower maximum on-current due to smaller normalized gate width compared to FinFET devices. Figure 5a, shows average SS values versus L_G for FinFETs with W_{FIN} = 40 nm. For L_G smaller than 40 nm, both linear and saturation SS increase substantially, due to short-channel effects. In Figure 5b, SS versus W_{FIN} for fixed L_G is plotted, showing that SS will benefit from further fin width scaling, due to improvement of the electrostatic control. The performance improvement achieved in this work compared to previously reported devices [15] are attributed to the use of an RMG scheme instead of a gate-first (GF) one. Here, the high-k/channel interface is less exposed to thermally induced degradation from the high-temperature RSD growth step, resulting in a lower density of interface traps.

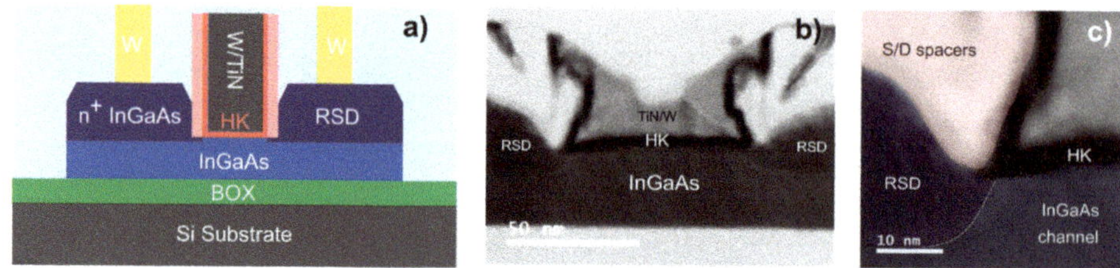

Figure 3. (**a**) STEM structural analysis. Cross-section schematic of completed InGaAs FET device, after final M1 metallization step. (**b**) STEM cross-section image for a device with LG = 60 nm. The RSD/channel interface is clearly distinguishable due to difference in indium content between the two. A false-colored zoomed view on the sidewall spacer/HK/RSD/channel interface is shown in (**c**).

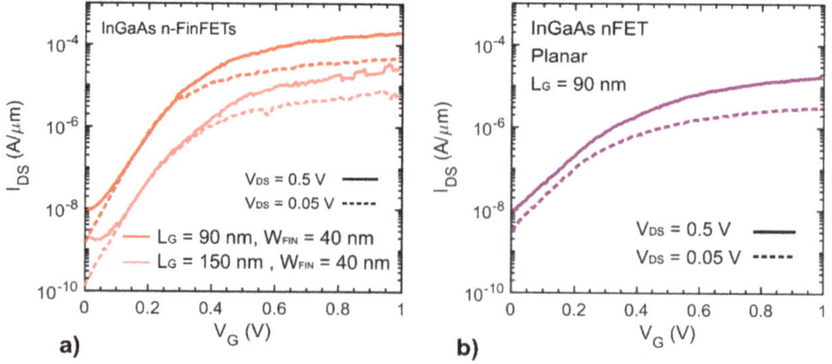

Figure 4. (**a**) Transfer characteristic of InGaAs FinFET device with L_G = 90 nm and L_G = 150 nm. Fin width is in both cases 40 nm. The shorter gate length device shows an on-current of 100 µA/µm and SS of 85 mV/decade. The gate leakage current (not shown) is at the limit of the measurement equipment, less than 1 pA. (**b**) Representative transfer characteristic of a planar InGaAs device with L_G = 90 nm.

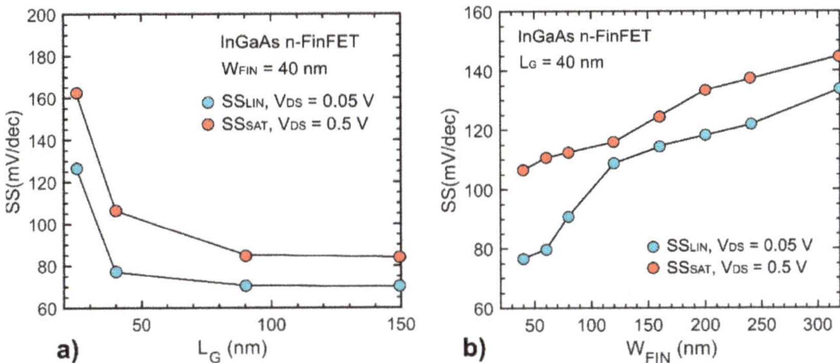

Figure 5. (**a**) Subthreshold slope values versus gate length for devices with W_{FIN} = 40 nm. For LG smaller than 40 nm, short channel effects start dominating the device behavior. (**b**) Average SS versus W_{FIN} for devices with L_G = 40 nm. The trend indicates that subthreshold performance benefits from further reducing fin width.

4. Conclusions

In this work, we have demonstrated InGaAs n-FinFETs devices using a novel III-V integration technique based on TASE. A full RMG process flow was developed including an improved RDS process with underlapping extensions. Devices with gate length of 90 nm and fin width of 40 nm show on-current of 100 µA/µm and subthreshold slope of about 85 mV/dec, demonstrating good electrostatic control. The strong off-state performance is due to the introduction of source-drain sidewall spacers combined with doped extensions achieved by digital etching. Results indicate that future performance improvements could be achieved by further scaling of gate length and fin width.

Author Contributions: H.S. performed the MOCVD InGaAs growth. C.C., C.Z., D.C., and L.C. fabricated the InGaAs FinFETs devices. C.C. and C.Z. performed the electrical characterization. K.M., H.S., and L.C. initiated the work on InGaAs epitaxy in oxide cavities. M.S. performed the STEM analysis and material characterization. C.C. conducted the electrical data analysis. C.C., with the support of all authors, wrote the manuscript.

Funding: This research was funded by INSIGHT grant number 688784 and REMINDER grant number 687931.

Acknowledgments: The authors want to acknowledge the BRNC staff for technical support.

Conflicts of Interest: The authors declare no conflict of interest.

References

1. Del Alamo, J.A. Nanometre-scale electronics with III–V compound semiconductors. *Nature* **2011**, *479*, 317–323. [CrossRef] [PubMed]
2. Hahn, H.; Deshpande, V.; Caruso, E.; Sant, S.; O'Connor, E.; Baumgartner, Y.; Sousa, M.; Caimi, D.; Olziersky, A.; Palestri, P.; et al. A scaled replacement metal gate InGaAs-on-Insulator n-FinFET on Si with record performance. In Proceedings of the 2017 IEEE International Electron Devices Meeting, San Francisco, CA, USA, 2–6 December 2017; pp. 17.5.1–17.5.4.
3. Zota, C.B.; Convertino, C.; Deshpande, V.; Merkle, T.; Sousa, M.; Caimi, D.; Czomomaz, L. InGaAs-on-Insulator MOSFETs Featuring Scaled Logic Devices and Record RF Performance. In Proceedings of the 2018 IEEE Symposium on VLSI Technology, Honolulu, HI, USA, 18–22 June 2018.
4. Sun, X.; D'Emic, C.; Cheng, C.-W.; Majumdar, A.; Sun, Y.; Cartier, E.; Bruce, R.L.; Frank, M.; Miyazoe, H.; Shiu, K.-T.; et al. High performance and Low Leakage Current InGaAs-on-Silicon FinFETs with 20 nm Gate Length. In Proceedings of the 2017 Symposia on VLSI Technology and Circuits, Kyoto, Japan, 5–8 June 2017; pp. 40–41.
5. Vardi, A.; del Alamo, J. Sub-10-nm fin-width self-aligned InGaAs FinFETs. *IEEE Electron Device Lett.* **2016**, *37*, 1104–1107. [CrossRef]

6. Czornomaz, L.; Daix, N.; Kerber, P.; Lister, K.; Caimi, D.; Rossel, C.; Sousa, M.; Uccelli, E.; Fompeyrine, J. Scalability of ultra-thin- body and BOX InGaAs MOSFETs on silicon. In Proceedings of the 2013 European Solid-State Device Research Conference, Bucharest, Romania, 16–20 September 2013; pp. 143–146.
7. Djara, V.; Deshpande, V.; Uccelli, E.; Daix, N.; Caimi, D.; Rossel, C.; Sousa, M.; Siegwart, H.; Marchiori, C.; Lubyshev, D.; et al. An InGaAs on Si platform for CMOS with 200 mm InGaAs-OI substrate, gate-first, replacement gate planar and FinFETs down to 120 nm contact pitch. In Proceedings of the 2015 Symposium on VLSI Technology (VLSI Technology), Kyoto, Japan, 16–18 June 2015; pp. T176–T177.
8. Convertino, C.; Zota, C.B.; Caimi, D.; Sousa, M.; Czornomaz, L. InGaAs FinFETs 3D Sequentially Integrated on FDSOI Si CMOS with Record Perfomance. In Proceedings of the 48th European Solid-State Device Research Conference (ESSDERC), Dresden, Germany, 3–6 September 2018; pp. 162–165.
9. Zota, C.B.; Lindelow, F.; Wernersson, L.E.; Lind, E. InGaAs tri-gate MOSFETs with record on-current. In Proceedings of the 2016 IEEE International Electron Devices Meeting (IEDM), San Francisco, CA, USA, 3–7 Decemebr 2016; pp. 3.2.1–3.2.4.
10. Lindelow, F. Gated Hall effect measurements on selectively grown InGaAs nanowires. *Nanotechnology* **2017**, *28*, 205204. [CrossRef] [PubMed]
11. Li, J.Z.; Bai, J.; Park, J.-S.; Adekore, B.; Fox, K.; Carroll, M.; Lochtefeld, A.; Shellenbarger, Z. Defect reduction of GaAs epitaxy on Si (001) using selective aspect ratio trapping. *Appl. Phys. Lett.* **2007**, *91*, 021114. [CrossRef]
12. Waldron, N.; Merckling, C.; Guo, W.; Ong, P.; Teugels, L.; I, S.A.; Tsvetanova, D.; Sebaai, F.; van Dorp, D.H.; Milenin, A.; et al. An InGaAs/InP quantum well finfet using the replacement fin process integrated in an RMG flow on 300 mm Si substrates. In Proceedings of the 2014 Symposia on VLSI Technology & Circuits, Honolulu, HI, USA, 9–13 June 2014; pp. 26–27.
13. Schmid, H.; Borg, M.; Moselund, K.; Gignac, L.; Breslin, C.M.; Bruley, J.; Cutaia, D.; Riel, H. Template-assisted selective epitaxy of III–V nanoscale devices for co-planar heterogeneous integration with Si. *Appl. Phys. Lett.* **2015**, *106*, 233101. [CrossRef]
14. Borg, M.; Schmid, H.; Moselund, K.E.; Cutaia, D.; Riel, H. Mechanisms of template-assisted selective epitaxy of InAs nanowires on Si. *J. Appl. Phys.* **2015**, *117*, 144303. [CrossRef]
15. Czornomaz, L.; Uccelli, E.; Sousa, M.; Deshpande, V.; Djara, V.; Caimi, D.; Rossell, M.D.; Erni, R.; Fompeyrine, J. Confined epitaxial lateral overgrowth (CELO): A novel concept for scalable integration of CMOS compatible InGaAs-on-insulator MOSFETs on large-area Si substrates. In Proceedings of the 2015 Symposium on VLSI Technology (VLSI Technology), Kyoto, Japan, 16–18 June 2015.
16. Cutaia, D.; Moselund, K.E.; Schmid, H.; Borg, M.; Olziersky, A.; Riel, H. Complementary III-V heterojunction lateral NW Tunnel FET technology on Si. In Proceedings of the 2016 IEEE Symposium on VLSI Technology, Honolulu, HI, USA, 14–16 June 2016; p. 403.
17. Convertino, C.; Zota, C.B.; Schmid, H.; Ionescu, A.M.; Moselund, K.E. III–V heterostructure tunnel field-effect transistor. *J. Phys. Condens. Matter* **2018**, *30*, 264005. [CrossRef] [PubMed]
18. Gooth, J.; Schaller, V.; Wirths, S.; Schmid, H.; Borg, M.; Bologna, N.; Karg, S.; Riel, H. Ballistic one-dimensional transport in InAs nanowires monolithically integrated on silicon. *Appl. Phys. Lett.* **2017**, *110*, 083105. [CrossRef]
19. Mauthe, S.; Mayer, B.; Sousa, M.; Villares, G.; Staudinger, P.; Schmid, H.; Moselund, K. Monolithically integrated InGaAs microdisk lasers on silicon using template-assisted selective epitaxy. *SPIE* **2018**. [CrossRef]
20. Borg, M.; Gignac, L.; Bruley, J.; Malmgren, A.; Sant, S.; Convertino, C.; Rossell, M.D.; Sousa, M.; Breslin, C.; Riel, H. Facet-selective group-III incorporation in InGaAs Template Assisted Selective Epitaxy. *IOP Nanotechnol.* **2018**. [CrossRef] [PubMed]

Article

Physics of Discrete Impurities under the Framework of Device Simulations for Nanostructure Devices

Nobuyuki Sano [1,*], Katsuhisa Yoshida [1], Chih-Wei Yao [2] and Hiroshi Watanabe [2]

1 Institute of Applied Physics, University of Tsukuba, Tsukuba, Ibaraki 305-8573, Japan; yoshida@bk.tsukuba.ac.jp

2 Department of Electrical and Computer Engineering, National Chiao Tung University, Hsinchu 30010, Taiwan; elegant.pegasus@gmail.com (C.-W.Y.); hwhpnabe@gmail.com (H.W.)

* Correspondence: sano.nobuyuki.gw@u.tsukuba.ac.jp; Tel.: +81-29-853-6479

Received: 21 November 2018; Accepted: 13 December 2018; Published: 16 December 2018

Abstract: Localized impurities doped in the semiconductor substrate of nanostructure devices play an essential role in understanding and resolving transport and variability issues in device characteristics. Modeling discrete impurities under the framework of device simulations is, therefore, an urgent need for reliable prediction of device performance via device simulations. In the present paper, we discuss the details of the physics associated with localized impurities in nanostructure devices, which are inherent, yet nontrivial, to any device simulation schemes: The physical interpretation and the role of electrostatic Coulomb potential in device simulations are clarified. We then show that a naive introduction of localized impurities into the Poisson equation leads to a logical inconsistency within the framework of the drift-diffusion simulations. We describe a systematic methodology for how to treat the Coulomb potential consistently with both the Poisson and current-continuity (transport) equations. The methodology is extended to the case of nanostructure devices so that the effects of the interface between different materials are taken into account.

Keywords: random dopant; drift-diffusion; variability; device simulation; nanodevice; screening; Coulomb interaction

---

1. Introduction

Although device miniaturization by following the traditional scaling rule has already ended, the pursuit of the scaling merit of Si-based electron devices is now directed toward utilizing three-dimensional gate-surrounding structures of the channel substrate and/or replacing the channel material by a new material such as Ge or compound semiconductors. Even atomic layers such as MoS_2 are also suggested as an alternative channel material [1]. Because of increasing complexity inherent to such advanced devices, the role of device simulation is getting more and more important [2]. In order to predict device characteristics accurately, it is essential to model physical phenomena based on the basic principles of physics. Local potential fluctuations induced by localized impurities, interface or line edge roughness, localized defects, etc., are just a few examples of such problems. Localized and, thus, discrete impurities doped in the device substrate induce surface potential fluctuations at the gate-oxide interface, which leads to threshold voltage fluctuations. This is called the random dopant fluctuations (RDFs) and a dominant factor that prevents further miniaturization of the present Si-based electron devices [3]. Intensive studies on the variability associated with discrete impurities have been, therefore, carried out in the past few decades [4–17]. The approaches employed in these studies scatter from the conventional drift-diffusion (DD) method to the Monte Carlo (MC) or the nonequilibrium Green's functions (NEGF) methods [18–23]. Most simulations are, however, somewhat empirical; discrete impurities are introduced into the Poisson equation as point charges or by simply replacing the atoms of the substrate with charged ions, and the variability in device characteristics has been evaluated by brute-force means.

We would like to stress that the physical modeling of such potential fluctuations under the framework of device simulations is not trivial. An introduction of localized impurities into the device simulations implies a transition from the conventional continuous (long-wavelength limit) picture, which is a primary assumption of all device simulations mentioned above, to the discontinuous (discrete) picture. In other words, a naive introduction of point charges or similar ones into the Poisson equation may lead to a logical inconsistency with self-consistently-coupled transport equations [7–9,24]. Nevertheless, almost no attention has been paid to the physical aspects of such discrete impurities, except the present authors' group. In the present paper, we thoroughly discuss the fundamental aspects of device modeling of randomly-doped discrete impurities within nanoscale device structures where the interface, as well as discreteness of impurities is of crucial importance. Although the physical issues are common to any kind of device simulations in which the Poisson equation is self-consistently coupled with the transport equations, we restrict our discussion here to the DD simulation scheme. A detailed analysis of the other simulation schemes along this line will be reported elsewhere.

The present paper is organized as follows. In Section 2, theoretical foundations imposed on the DD device simulations are discussed with emphasis on the length-scale involved in the scheme. In Section 3, the physics behind discrete impurities in the bulk is discussed, and a discrete impurity model appropriate for nanostructures where the interface effect between two different materials on the potential is inevitable is proposed. Finally, conclusions are drawn in Section 4.

2. Theoretical Foundations of Drift-Diffusion Device Simulations

The physical origin of RDFs in Si-MOSFETs has been properly recognized from the early stage of the investigations [7–9]; it is the long-range part of the Coulomb potential of doped impurities in the device substrate whose fluctuations lead to the variability of device characteristics. Since the potential fluctuations result from the discreteness of impurities, it is inevitable to introduce the discrete nature of impurities into the Poisson equation. Since the impurity density is described by a continuous and smooth function in the conventional scheme, this is not an easy task. If a point charge is introduced into the Poisson equation, the resulting Coulomb potential becomes so steep that carriers with opposite charge are trapped by the attractive potential (It should be noted that trapping and detrapping processes of carriers by ionized impurities are caused through the bound states created by the impurity Coulomb potential. These bound states, sometimes disguised in the DD simulations by the effective quantum potential by the density gradient method, are totally different from the free traveling states of carriers discussed here. This point should not be confused, as often seen in the literature, with the present issue.), and the doping density in the substrate is effectively lowered, which leads to an artificial threshold voltage shift [8,9]. Furthermore, as we shall discuss in Section 2.1, the short-range part of the Coulomb potential of impurities is double-counted because the conventional mobility model employed in DD simulations is dependent on impurity density, and thus, impurity scattering induced by the short-range (screened) Coulomb potential is already taken into account through the mobility model in the current-continuity equation [7–9,24,25]. A key to resolve this problem lies in the fact that we must treat the length-scales presumed in both the Poisson and the transport (current-continuity) equations in a consistent way.

2.1. Meaning of the Long-Wavelength Limit in Drift-Diffusion Simulations

The DD simulation scheme consists of the following Poisson equation,

$$\nabla \cdot [\varepsilon_s \nabla \phi (\mathbf{r}, t)] = -e \left\{ p (\mathbf{r}, t) - n (\mathbf{r}, t) + N_d^+ (\mathbf{r}) - N_a^- (\mathbf{r}) \right\}, \tag{1}$$

and the current-continuity equation for electrons,

$$\frac{\partial n(\mathbf{r},t)}{\partial t}-\frac{1}{e}\nabla\cdot\mathbf{J}_n(\mathbf{r},t)=G_n(\mathbf{r},t)-R_n(\mathbf{r},t). \tag{2}$$

Here, ϕ is the electric potential, e (>0) is the magnitude of the electron charge, n and p are electron and hole densities, ε_s is the dielectric constant, N_d^+ and N_a^- are ionized donor and acceptor densities, and G_n and R_n are generation and recombination rates per unit time. The current density of electrons is given by the sum of drift and diffusion current densities,

$$\mathbf{J}_n(\mathbf{r},t)=en(\mathbf{r},t)\,\mu_n(-\nabla\phi(\mathbf{r},t))-eD_n(-\nabla n(\mathbf{r},t)). \tag{3}$$

where μ_n and D_n are the mobility and diffusion constants of electrons, respectively. In addition, similar equations for holes are coupled with the above equations to determine the hole density. Notice that the current-continuity equation plays the role of the transport equation under the framework of the DD device simulations in the sense that the current-continuity equation determines carrier density (the Boltzmann transport equation and the Keldysh equation play the role of transport equations in the MC and NEGF device simulation schemes, respectively).

The Poisson equation given by Equation (1) holds true at any length-scale. Namely, if the charge density on the right-hand side is expressed in terms of the delta-functions (point charges), then the potential profile contains all wavelengths with no bounds. However, the potential usually assumed in Equation (1) is the one under the "long-wavelength limit" in the conventional device simulations so that the charge density of Equation (1) is expressed by a smooth continuous function (jelly impurity). This requirement is consistent with the current continuity Equation (3); the first term $\mu_n(-\nabla\phi)$ represents not the thermal velocity, but the drift velocity, which results from the collective (averaged) motion of electrons, and the second term yields the diffusion current, which results from the gradient of a smooth continuous electron density. Therefore, the DD simulation scheme is indeed consistent with respect to the length-scale as far as all physical variables are expressed with those under the long-wavelength limit.

The mathematical meaning of the "long-wavelength limit" is interpreted as follows. Let us consider the microscopic impurity density N_{micro} expressed by the delta functions such that:

$$N_{micro}(\mathbf{r})=\sum_{i=1}^{N_{imp}}\delta(\mathbf{r}-\mathbf{R}_i)=\frac{N_{imp}}{\Delta V}+\sum_{i=1}^{N_{imp}}\frac{1}{\Delta V}\sum_{\mathbf{k}\neq0}e^{i\mathbf{k}\cdot(\mathbf{r}-\mathbf{R}_i)}, \tag{4}$$

where N_{imp} is the number of impurities included in a small volume ΔV around the position $\mathbf{r}$ and $\mathbf{R}_i$ is the position of the ith impurity. Then, the macroscopic (jelly) impurity density $\bar{N}(\mathbf{r})$, i.e., the density under the long-wavelength limit, is given by averaging $N_{micro}(\mathbf{r})$ over the small volume ΔV,

$$\bar{N}(\mathbf{r})=\frac{1}{\Delta V}\int_{\Delta V}d^3\mathbf{r}\,N_{micro}(\mathbf{r})=\frac{N_{imp}}{\Delta V}. \tag{5}$$

This implies that the long-wavelength limit of impurity density is equivalent to taking account of only the zero-Fourier component of the microscopic impurity density given by Equation (4). In other words, when impurity density is expressed as a smooth continuous function of position, the impurity density is considered to be locally flat, so that the electrostatic potential induced by the impurities is also locally flat because the number of impurities included in the region ΔV is virtually regarded as constant. This situation is schematically shown in Figure 1.

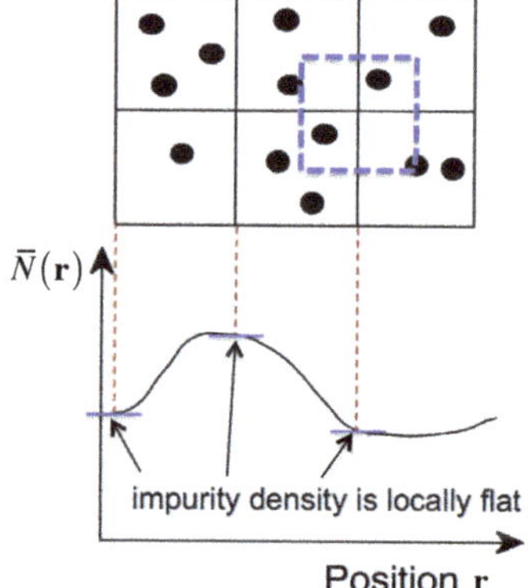

Figure 1. Schematic drawing of the spatial configuration of discrete impurities (above) and the corresponding macroscopic impurity density under the long-wavelength limit (below). The macroscopic density assumes that the number of impurities included in the neighborhood is virtually constant and the electrostatic potential is also virtually flat.

The question arises why non-zero Fourier components of discrete impurity density (equivalently, of electrostatic potential induced by each discrete impurity) could be ignored in the Poisson Equation (1). The answer lies in the fact that the mobility employed in Equation (3) is usually modeled as a function of impurity density in DD simulations, as already mentioned above. Since the mobility is determined by scattering, the impurity density dependence of mobility implies that non-zero Fourier components of each impurity potential are regarded as scattering potential and included in the conventional mobility model in Equation (3). This is the reason why non-zero Fourier components of the discrete impurity density and, thus, of the impurity potential are eliminated in the Poisson equation. Otherwise, these Fourier components would be double-counted.

2.2. Incomplete Screening of the Long-Range Part of the Coulomb Potential

We would like to stress, however, that all non-zero Fourier components of the Coulomb potential of each impurity are not actually treated as scattering potential. Impurity-limited mobility μ or scattering time τ appearing in the formula of mobility, $\mu = e\tau/m^*$ (m^*: effective mass), is, in most cases, calculated with the screened Yukawa potential. In other words, it is the short-range part of the Coulomb potential with the wavelength shorter than the screening length λ_c that is treated as the scattering potential. The non-zero Fourier components with the wavelength larger than λ_c, which is hereafter denoted as the long-range part of the Coulomb potential, is assumed to be completely canceled by the induced charges of screening carriers and, thus, ignored in both the Poisson equation and the current-continuity equations.

It should be noted that the above scenario holds true if the carrier density is nearly equal to or above the average impurity density. The device is, however, operated under the extreme nonequilibrium conditions, and the charge neutrality is also broken in the subthreshold regimes near the gate interface where carrier density is very small. As a result, ionized impurities are not completely screened by carriers, and some portion of the long-range part of the Coulomb potential is left unscreened and appears as potential fluctuations (on the other hand, the impurities are "over-"screened in the inversion regimes by carriers whose density could be larger than the impurity density. In this case, however, carriers do not see the charge of impurities once the carrier density exceeds the impurity density, and thus, such "over-screening" does not induce any long-range potential fluctuations.). This unscreened portion of the potential always exists, no matter how large the device is. The reason why such potential fluctuations are not observable in large devices is because the variability of device characteristics associated with potential fluctuations is self-averaged due to large number of impurities in the substrate [26,27]. As the device shrinks, self-averaging is no longer strong enough to suppress the fluctuations and the variability in device properties such that RDF becomes significant. Therefore, the physical origin of RDF is indeed the long-range part of the Coulomb potential resulting from incomplete screening, as conjectured in the previous studies [7–9].

Furthermore, we should notice that as the volume of the channel substrate is small and/or complicated such as fin or surrounding-gate (nanowire) structures, the boundary (interface) greatly affects the spatial distribution of induced charges for screening in the semiconductor substrate. In other words, the effect of interfaces needs to be taken into account properly to extract the long-range part of the Coulomb potential.

3. Discrete Impurity Models for Drift-Diffusion Simulation

Following the arguments in Section 2, we need to somehow introduce the long-range part of the Coulomb potential associated with incomplete screening of discrete impurities. A critical issue is how we could include such potential fluctuations within the framework of the DD simulations. The hint is provided by recalling the role of the long-range part of the Coulomb potential. The most obvious example is plasma wave excitations in electron gas; electrons tend to screen the external (or extra) potential disturbance, yet because of the inertia of electrons, electron density spatially exceeds or gets under the proper value of the density, and this leads to plasma oscillations. This phenomena exactly corresponds to the dynamical version of the complete and incomplete screening situations described in Section 2.2. Since the plasma oscillation results from the collective motions of electrons, it is natural to include the long-range part of the Coulomb potential as the self-consistent Hartree potential in the Poisson equation, not as scattering in the mobility model of the current-continuity equation. Along with this idea, we have previously proposed a discrete impurity model for DD simulations [7], in which the charge density of each impurity in the Poisson equation is spread over the screening length so that the short-range part of the Coulomb potential is eliminated from the self-consistent Hartree potential (Technically, this may look similar to the cloud-in-cell method used in the MC simulations. However, the concept behind this process is very different. In the cloud-in-cell method, the size of charged particle is dependent on the mesh employed in the simulations, whereas the size of charged particle is fixed in the present method with the screening length and, thus, independent of the mesh.).

3.1. Discrete Impurity in the Bulk Structure

We notice that the long-range part of the Coulomb potential is just the potential (except the sign of charge polarity) generated by induced charges to screen ionized impurities. Since the self-consistent potential ϕ_{sc} is given by the sum of the external (impurity) potential ϕ_{ext} and the induced potential $\delta\phi$, the long-range part of the impurity Coulomb potential ϕ_l, which results from the Poisson equation under the framework of DD simulations, is obtained from:

$$\phi_l(\mathbf{r}) = -\delta\phi(\mathbf{r}) = -\{\phi_{sc}(\mathbf{r}) - \phi_{ext}(\mathbf{r})\}. \tag{6}$$

Transforming it into the Fourier $\mathbf{q}$-space and using the fact that $\phi_{sc}(\mathbf{q}) = \phi_{ext}(\mathbf{q})/\varepsilon(\mathbf{q})$ with the (relative) static dielectric function $\varepsilon(\mathbf{q})$, Equation (6) is expressed as:

$$\phi_l(\mathbf{q}) = \left(1 - \frac{1}{\varepsilon(\mathbf{q})}\right)\phi_{ext}(\mathbf{q}) \tag{7}$$

and the corresponding charge distribution $\rho_l(\mathbf{q})$ is given by:

$$\rho_l(\mathbf{q}) = \varepsilon_s q^2 \left(1 - \frac{1}{\varepsilon(\mathbf{q})}\right)\phi_{ext}(\mathbf{q}), \tag{8}$$

where $\varepsilon(\mathbf{q})$ is calculated by the self-consistent field (random phase) approximation [28] and given by:

$$\varepsilon(\mathbf{q}) = 1 + \frac{{q_c}^2}{q^2} F(\mathbf{q}). \tag{9}$$

Here, q_c is the inverse of the Debye screening length and given by $q_c = \sqrt{\bar{n}(\mathbf{r})\, e^2 / (\varepsilon_s k_B T)}$ with average carrier density $\bar{n}(\mathbf{r})$ at equilibrium (The interpretation of $\bar{n}(\mathbf{r})$ requires some care. From the arguments in Section 2, it should be interpreted as the carrier density at equilibrium under the condition that charge neutrality is preserved. Here, charge neutrality simply means that the (macroscopic) carrier density is equal to the background (macroscopic) dopant density. Thus, $\bar{n}(\mathbf{r})$ should take the same value as that at the flat-band condition even if actual carrier density is different in the depletion or inversion regimes.), k_B is the Boltzmann constant, and T is temperature. Assuming that the carrier distribution function is approximated by the classical Boltzmann statistics, $F(\mathbf{q})$ is given by:

$$F(\mathbf{q}) = -\frac{\sqrt{2m^* k_B T}}{\hbar q} \operatorname{Re}\left[Z\left(\frac{\hbar q}{2\sqrt{2m^* k_B T}} \right) \right], \tag{10}$$

where m^* is the effective mass of carriers in the semiconductor, $\hbar$ is the Planck constant divided by 2π, and $Z(\alpha)$ is the plasma dispersion function defined by:

$$Z(\alpha) = \lim_{\delta \to 0+} \frac{1}{\sqrt{\pi}} \int_{-\infty}^{\infty} d\xi \frac{e^{-\xi^2}}{\xi - \alpha - i\delta}. \tag{11}$$

Since we are concerned with the long-range part of the potential, $F(\mathbf{q})$ is, as usual, set to unity, and the dielectric function could be approximated by the well-known Thomas–Fermi expression.

Consequently, the long-range part of the impurity potentials ϕ_l in both $\mathbf{q}$- and $\mathbf{r}$-spaces is given by:

$$\phi_l(\mathbf{q}) = \frac{e}{\varepsilon_s} \frac{{q_c}^2}{q^2 \left(q^2 + {q_c}^2\right)} \tag{12}$$

and:

$$\phi_l(\mathbf{r}) = \frac{e}{4\pi\varepsilon_s} \left(\frac{1}{r} - \frac{e^{-q_c r}}{r} \right), \tag{13}$$

respectively. As expected, $\phi_l(\mathbf{r})$ is indeed the difference between the bare Coulomb potential and the short-range Yukawa potential. The corresponding charge distributions ρ_l in $\mathbf{q}$- and $\mathbf{r}$-spaces are given by:

$$\rho_l(\mathbf{q}) = e \frac{{q_c}^2}{q^2 + {q_c}^2} \tag{14}$$

and:

$$\rho_l(\mathbf{r}) = e \frac{{q_c}^2}{4\pi} \frac{e^{-q_c r}}{r}, \tag{15}$$

respectively. Notice that it is this $\rho_l(\mathbf{r})$ that should replace the charge density expressed by a point charge of discrete impurity in the Poisson equation. Then, we are able to extract the long-range portion from the bare Coulomb potential. We would like to stress again that this long-range potential appears as potential fluctuation when the charge neutrality condition in the substrate is broken so that the screening by carriers is incomplete. In other words, the screening effect usually suppresses potential fluctuation, and thereby, incomplete screening causes potential fluctuation.

It is very interesting to compare Equations (14) and (15) with those we have previously proposed to extract the long-range portion of the impurity Coulomb potential [7–9,25]. Figure 2 shows the charge distributions (in both **q**- and **r**-spaces) and the (long-range) potentials of the three different expressions; the Yukawa-like expression given by Equation (15), the long-range expression given in [7–9], and the Gaussian expression in [25] employed as an alternative model to eliminate artificial oscillations showing up in the charge density by the previous long-range model. Explicit formulas of all three models are also shown in the inset of Figure 2. It is clear that the only difference among the models is how the short-range part of the Coulomb potential is eliminated: the short-range part is sharply cut at q_c in the previous long-range model, whereas the present and Gaussian models gradually eliminate the short-range part. As a result, the long-range potentials are slightly different near the origin. Yet, all models properly approach the bare Coulomb potential as the distance from the impurity becomes much greater than $1/q_c$.

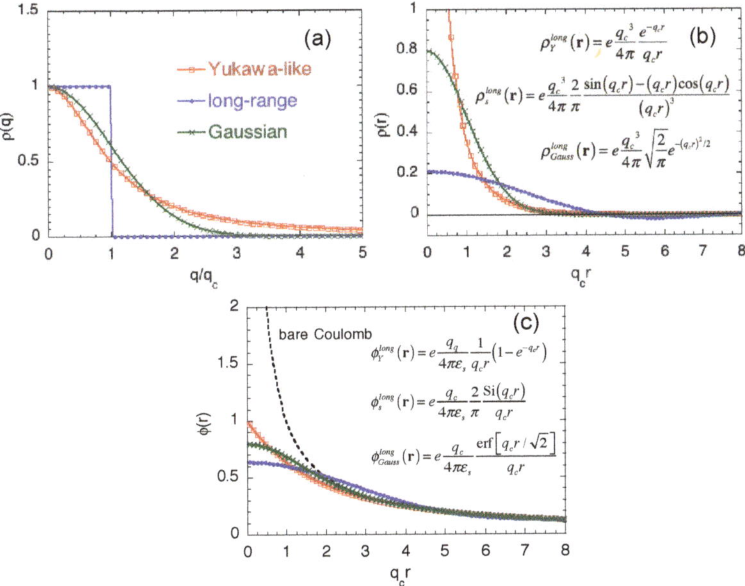

Figure 2. Charge densities of the discrete impurity located at origin as a function of (**a**) normalized wave-number q/q_c and (**b**) normalized position $q_c r$ for three different discrete impurity models, Yukawa-like (red), long-range (blue), and Gaussian (green). The charge densities in **q**-space and **r**-space are, respectively, normalized by e and $eq_c^3/4\pi$. (**c**) Long-range potential of the three discrete impurity models as a function of normalized position $q_c r$. The bare Coulomb potential is also shown with the black dotted curve. The potential is normalized by $eq_c/4\pi\varepsilon_s$.

It should be noted that all expressions of the charge distribution hold true only in bulk structures because they assume that screening is not disturbed by the boundaries. This situation breaks down in nanostructures, in which impurities are surrounded by an interface and/or boundary.

3.2. Discrete Impurity Including the Effects of Interface

We now extend the above discrete impurity model to the case where an impurity is located near interface of two different materials so that the boundary could modulate the long-range part of the impurity potential. In order to extract the long-range part of the Coulomb potential, we take a similar methodology to the one we have employed in Section 3.1.

Let us consider two different materials with the relative permittivities ε_1 and ε_2 that are separated by an infinite plane interface. An ionized impurity is then embedded in the material with ε_1 at a distance a (>0) from the interface. The cylindrical coordinates whose origin coincides with the position of the impurity are employed, as shown in Figure 3.

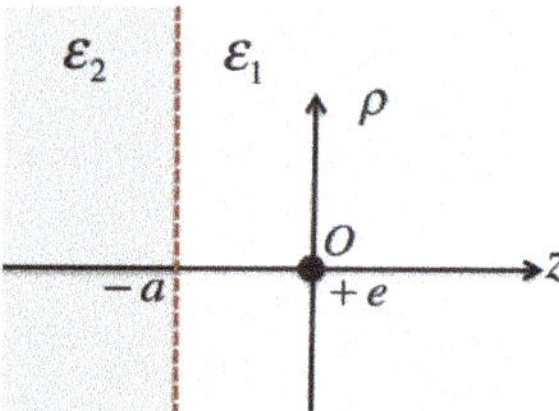

Figure 3. Schematic drawing of the interface between the two different materials with ε_1 and ε_2. A point-charge impurity is placed at the origin that is a distance a from the interface. The cylindrical coordinates are employed for the calculations.

The external (impurity) potential ϕ_{ext} that satisfies the boundary condition at the interface is expressed as:

$$\phi_{ext}(\rho, z) = \frac{e}{4\pi\varepsilon_1}\int_0^\infty dk\left[e^{-k|z|} + \alpha e^{-k(z+2a)}\right] J_0(k\rho) \tag{16}$$

for $z \geq -a$ and:

$$\phi_{ext}(\rho, z) = \frac{e}{4\pi\varepsilon_1}\int_0^\infty dk\,(1+\alpha)\, e^{kz} J_0(k\rho) \tag{17}$$

for $z \leq -a$. Here, $\alpha = (\varepsilon_1 - \varepsilon_2)/(\varepsilon_1 + \varepsilon_2)$, and $J_0(x)$ is the zeroth order Bessel function. The Fourier transform of $\phi_{ext}(\rho, z)$ is then given by:

$$\phi_{ext}(\mathbf{q}) = \frac{e}{\varepsilon_1}\frac{1}{q_\perp{}^2 + q_z{}^2}\left(1 + \alpha e^{-q_\perp a + i q_z a}\right) \equiv \frac{e}{\varepsilon_1}\frac{1}{q^2}\{1 + \gamma(\mathbf{q})\}. \tag{18}$$

Here, $\mathbf{q} = (q_\perp, q_z)$, where $q_\perp$ and q_z are, respectively, wavenumbers normal to and along the z-axis so that $q^2 = q_\perp^2 + q_z^2$.

The static dielectric function $\varepsilon(\mathbf{q})$ under the self-consistent approximation becomes

$$\varepsilon(\mathbf{q}) = 1 + \frac{q_c{}^2}{q^2}\{1 + \gamma(\mathbf{q})\} F(\mathbf{q}). \tag{19}$$

Noting that $|\gamma(\mathbf{q})| = \left|\alpha e^{-q_\perp a + i q_z a}\right| < 1$ and $0 < F(\mathbf{q}) \leq 1$, we employ the same approximation as in the bulk case; $\varepsilon(\mathbf{q})$ is approximated by the simple Thomas–Fermi expression. Hence, we can write the charge distribution induced by the impurity (with opposite charge polarity) as:

$$\rho_I(\mathbf{q}) = \varepsilon_s q^2\left(1 - \frac{1}{\varepsilon(\mathbf{q})}\right)\phi_{ext}(\mathbf{q}) = e\frac{q_c{}^2}{q^2 + q_c{}^2}\{1 + \gamma(\mathbf{q})\}. \tag{20}$$

Transforming it into the r-space, we obtain the charge distribution $\rho_I(\rho,\ z)$ by:

$$\rho_I(\rho,\ z) = e\frac{q_c{}^2}{4\pi}\int_0^\infty dq_\perp \frac{q_\perp}{\sqrt{q_\perp{}^2 + q_c{}^2}}\left[e^{-|z|\sqrt{q_\perp{}^2 + q_c{}^2}} + \alpha e^{-(z+a)\sqrt{q_\perp{}^2 + q_c{}^2} - q_\perp a}\right] J_0(q_\perp \rho). \tag{21}$$

Notice that this expression is valid only in the region of $z \geq -a$.

Figure 4 shows the charge distributions given by Equation (21) along the z-axis ($\rho = 0$), which are induced by the impurity at a distance a from the interface. The charge distributions of three different distances from the interface are shown for two different oxides in $z < -a$; SiO_2 ($\varepsilon_2 = 3.9$) and

HfO_2 ($\varepsilon_2 = 25$). The semiconductor material in $z > -a$ where the impurity resides is chosen to be Si ($\varepsilon_1 = 11.8$). In Figure 4, the charge distributions given by the first and second terms of the integrand in Equation (21) are also shown.

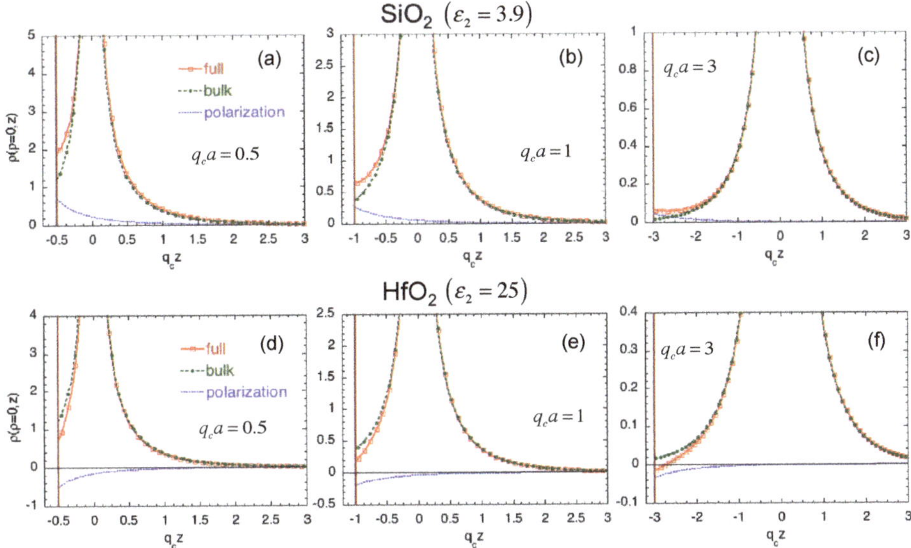

Figure 4. Charge distributions (denoted by "full" with red line) along the z-axis ($\rho = 0$) induced by the impurity at the origin for two different dielectrics; SiO_2 with $\varepsilon_2 = 3.9$ (**a–c**) and HfO_2 with $\varepsilon_2 = 25$ (**d–f**). The distances from the interface are $q_c a = 0.5$ (**a**,**d**), 1 (**b**,**e**), and 3 (**c**,**f**). The semiconductor material in $z > -a$ is assumed to be Si ($\varepsilon_1 = 11.8$). The charge distributions obtained from the first term (denoted by "bulk" with the green line) and the second term (denoted by "polarization" with the blue line) in the integrand of Equation (21) are also shown.

We notice that the first term in Equation (21) is identical to the charge distribution given by Equation (15). Therefore, this term represents the charge distribution induced by the impurity itself and should be interpreted similarly to the case of the bulk. The second term results from the polarization charge at the interface at $z = -a$ and appears only after the discreteness of impurity is taken into account. Furthermore, depending on the magnitude of ε_2 compared with ε_1, the polarity of induced polarization charge changes: It is positive for SiO_2, whereas it is negative for HfO_2. As a result, the total charge distribution is more heavily affected near the interface. It should be pointed out that Equation (21) is derived by ignoring the metal surface on top of the oxide layer. Strictly speaking, this effect should be also taken into account if the oxide thickness is very small. This might be the case of SiO_2. However, the effect is negligible for the case of HfO_2 because of large permittivity (and large thickness usually employed in reality).

In order to properly take into account the polarization at the interface of discrete impurity in DD simulations, the following methodology is suggested. Since the first term in Equation (21) is identical to Equation (15), an impurity charge should spread similarly to the case of the bulk in accordance with Equation (15). It should be noted, however, that the charge distribution extending over the other side of the material ($z < -a$ in the present case) should fold back to the semiconductor side ($z > -a$). Otherwise, the impurity density within the semiconductor substrate would not be conserved. This is exactly the procedure we have taken in the previous models when an impurity is located near interface so that some portion of its charge distribution spreads over the oxide. In addition, the correction charge distribution $\delta\rho_l$ associated with the polarization at the interface needs to be included. As a result, the impurity charge density near the interface for DD simulations should be given by:

$$\rho_l^{DD}(\rho, z) = \rho_l^{bulk}(\rho, z) + \delta\rho_l(\rho, z)$$
$$= e\frac{q_c{}^2}{4\pi}\frac{e^{-q_c\sqrt{\rho^2+z^2}}}{\sqrt{\rho^2+z^2}} + e\frac{q_c{}^2}{4\pi}\alpha\int_0^\infty dq_\perp \frac{q_\perp}{\sqrt{q_\perp{}^2+q_c{}^2}} e^{-(z+a)\sqrt{q_\perp{}^2+q_c{}^2}-q_\perp a} J_0(q_\perp\rho) \quad (22)$$

where ρ_l^{bulk} is evaluated over the entire region, namely the charge distribution extended beyond the interface ($z < -a$) is folded back to the semiconductor substrate ($z > -a$) symmetrically with respect to the interface, whereas $\delta\rho_l$ is evaluated only in the semiconductor substrate ($z > -a$) (we are currently applying for a patent on a more tractable method, which could be properly implemented in the DD simulators, to extract the long-range potential of the discrete impurity near the interface). Clearly, Equation (22) coincides with the discrete impurity model in the bulk as the impurity position becomes very far from the interface (The present model is applicable to the region where impurities are intentionally doped so that the macroscopic dopant density is well defined in the device substrate. The case in which impurities are not intentionally doped is out of the scope of the present analysis.).

4. Conclusions

We have discussed the details of the physics associated with localized impurities in nanostructure devices. The physical interpretation and the role of the electrostatic Coulomb potential of localized impurities under the framework of device simulations have been clarified. We have shown that a naive introduction of localized impurities into the Poisson equation leads to a logical inconsistency within the scheme of the DD simulation. We have developed a systematic methodology for how to treat the Coulomb potential consistently with both the Poisson and the current-continuity (transport) equations. We have demonstrated that this method naturally leads to the concept of the long-range discrete impurity model we have proposed before. The method has been extended to the case of nanostructure devices in which the effects of the interface between different materials are taken into account.

Finally, we would like to point out that the present analysis is also closely related to the treatment of the Coulomb potential in any device simulation schemes. The long-wavelength limit is usually the common assumption in the Poisson equation of most device simulations, and thus, a similar careful analysis on logical consistency between the Poisson and the transport equations is required. This issue is under progress and will be reported elsewhere.

Author Contributions: Conceptualization and methodology, N.S.; validation and analysis, N.S., K.Y., C.-W.Y., H.W.; writing, original draft preparation, N.S.; writing, review and editing, N.S., K.Y., C.-W.Y. and H.W.

Funding: This research received no external funding.

Conflicts of Interest: The authors declare no conflict of interest.

References

1. Fiori, G.; Bonaccorso, F.; Iannaccone, G.; Palacios, T.; Neumaier, D.; Seabaugh, A.; Banerjee, S.K.; Colombo, L. Electronics based on two-dimensional materials. *Nat. Nanotechnol.* **2014**, *9*, 768–779. [CrossRef] [PubMed]
2. Zographos, N.; Zechner, C.; Martin-Bragado, I.; Lee, K.; Oh, Y.S. Multiscale modeling of doping processes in advanced semiconductor devices. *Mater. Sci. Semicond. Process.* **2017**, *62*, 49–61. [CrossRef]
3. Tsunomura, T.; Nishida, A.; Hiramoto, T. Verification of Threshold Voltage Variation of Scaled Transistors with Ultralarge-Scale Device Matrix Array Test Element Group. *Jpn. J. Appl. Phys.* **2009**, *48*, 124505. [CrossRef]
4. Nishinohara, K.; Shigyo, N.; Wada, T. Effects of microscopic fluctuations in dopant distributions on MOSFET threshold voltage. *IEEE Trans. Electron Dev.* **1992**, *39*, 634–639. [CrossRef]

5. Wong, H.S.; Taur, Y. Three-dimensional "atomistic" simulation of discrete random dopant distribution effects in sub-0.1 μm MOSFET's. In Proceedings of the IEEE International Electron Devices Meeting, Technical Digest, Washington, DC, USA, 5–8 December 1993; pp. 705–708.
6. Stolk, P.A.; Klaassen, D.B.M. The effect of statistical dopant fluctuations on MOS device performance. In Proceedings of the International Electron Devices Meeting, Technical Digest, San Francisco, CA, USA, 8–11 December 1996; pp. 627–630.
7. Sano, N.; Matsuzawa, K.; Mukai, M.; Nakayama, N. Role of long-range and short-range Coulomb potentials in threshold characteristics under discrete dopants in sub-0.1 μm Si-MOSFETs. In Proceedings of the International Electron Devices Meeting, Technical Digest, San Francisco, CA, USA, 11–13 December 2000; pp. 275–278.
8. Sano, N.; Tomizawa, M. Random dopant model for three-dimensional drift-diffusion simulations in metal-oxide-semiconductor field-effect-transistors. *Appl. Phys. Lett.* **2001**, *79*, 2267–2269. [CrossRef]
9. Sano, N.; Matsuzawa, K.; Mukai, M.; Nakayama, N. On discrete random dopant modeling in drift-diffusion simulations: physical meaning of 'atomistic' dopants. *Microelectron. Reliab.* **2002**, *42*, 189–199. [CrossRef]
10. Asenov, A.; Brown, A.R.; Davies, J.H.; Kaya, S.; Slavcheva, G. Simulation of intrinsic parameter fluctuations in decananometer and nanometer-scale MOSFETs. *IEEE Trans. Electron Dev.* **2003**, *50*, 1837–1852. [CrossRef]
11. Damrongplasit, N.; Shin, C.; Kim, S.H.; Vega, R.A.; Liu, T.K. Study of Random Dopant Fluctuation Effects in Germanium-Source Tunnel FETs. *IEEE Trans. Electron Dev.* **2011**, *58*, 3541–3548. [CrossRef]
12. Damrongplasit, N.; Kim, S.H.; Liu, T.K. Study of Random Dopant Fluctuation Induced Variability in the Raised-Ge-Source TFET. *IEEE Electron Dev. Lett.* **2013**, *34*, 184–186. [CrossRef]
13. Li, Y.; Chang, H.; Lai, C.; Chao, P.; Chen, C. Process variation effect, metal-gate work-function fluctuation and random dopant fluctuation of 10-nm gate-all-around silicon nanowire MOSFET devices. In Proceedings of the International Electron Devices Meeting, Technical Digest, Washington, DC, USA, 7–9 December 2015; pp. 34.4.1–34.4.4.
14. Yoon, J.S.; Rim, T.; Kim, J.; Kim, K.; Baek, C.K.; Jeong, Y.H. Statistical variability study of random dopant fluctuation on gate-all-around inversion-mode silicon nanowire field-effect transistors. *Appl. Phys. Lett.* **2015**, *106*, 103507. [CrossRef]
15. Chen, C.Y.; Lin, J.T.; Chiang, M.H. Threshold-voltage variability analysis and modeling for junctionless double-gate transistors. *Microelectron. Reliab.* **2017**, *74*, 22–26. [CrossRef]
16. Yu, H.; Kim, D.; Rhee, S.; Choi, S.; Park, Y.J. A Mobility Model for Random Discrete Dopants and Application to the Current Drivability of DRAM Cell. *IEEE Trans. Electron Dev.* **2017**, *64*, 4246–4251. [CrossRef]
17. Yoon, J.; Baek, R. Study on Random Dopant Fluctuation in Core Shell Tunneling Field-Effect Transistors. *IEEE Trans. Electron Dev.* **2018**, *65*, 3131–3135. [CrossRef]
18. Dollfus, P.; Bournel, A.; Galdin, S.; Barraud, S.; Hesto, P. Effect of discrete impurities on electron transport in ultrashort MOSFET using 3D MC simulation. *IEEE Trans. Electron Dev.* **2004**, *51*, 749–756. [CrossRef]
19. Alexander, C.; Roy, G.; Asenov, A. Random-Dopant-Induced Drain Current Variation in Nano-MOSFETs: A Three-Dimensional Self-Consistent Monte Carlo Simulation Study Using "Ab Initio" Ionized Impurity Scattering. *IEEE Trans. Electron Dev.* **2008**, *55*, 3251–3258. [CrossRef]
20. Martinez, A.; Aldegunde, M.; Seoane, N.; Brown, A.; Barker, J.; Asenov, A. Quantum-Transport Study on the Impact of Channel Length and Cross Sections on Variability Induced by Random Discrete Dopants in Narrow Gate-All-Around Silicon Nanowire Transistors. *IEEE Trans. Electron Dev.* **2011**, *58*, 2209–2217. [CrossRef]
21. Georgiev, V.P.; Towie, E.A.; Asenov, A. Impact of Precisely Positioned Dopants on the Performance of an Ultimate Silicon Nanowire Transistor: A Full Three-Dimensional NEGF Simulation Study. *IEEE Trans. Electron Dev.* **2013**, *60*, 965–971. [CrossRef]
22. Sellier, J.; Dimov, I. The Wigner-Boltzmann Monte Carlo method applied to electron transport in the presence of a single dopant. *Compt. Phys. Commun.* **2014**, *185*, 2427–2435. [CrossRef]
23. Carrillo-Nuñez, H.; Lee, J.; Berrada, S.; Medina-Bailón, C.; Adamu-Lema, F.; Luisier, M.; Asenov, A.; Georgiev, V.P. Random Dopant-Induced Variability in Si-InAs Nanowire Tunnel FETs: A Quantum Transport Simulation Study. *IEEE Electron Dev. Lett.* **2018**, *39*, 1473–1476. [CrossRef]
24. Sano, N. Physical issues in device modeling: Length-scale, disorder, and phase interference. In Proceedings of the International Conference on Simulation of Semiconductor Processes and Devices (SISPAD), Kamakura, Japan, 7–9 September 2017; pp. 1–4.

25. Karasawa, T.; Nakanishi, K.; Sano, N. Discrete impurity and mobility in drift-diffusion simulations for device characteristics variability. In Proceedings of the International Semiconductor Device Research Symposium, College Park, MD, USA, 9–11 December 2009; pp. 1–2.
26. Sano, N. Impurity-limited resistance and phase interference of localized impurities under quasi-one dimensional nano-structures. *J. Appl. Phys.* **2015**, *118*, 244302. [CrossRef]
27. Sano, N. Variability and self-average of impurity-limited resistance in quasi-one dimensional nanowires. *Solid-State Electron.* **2017**, *128*, 25–30. [CrossRef]
28. Kittel, C. *Quantum Theory of Solids*, 2nd Rev. ed.; John-Wiley & Sons: Hoboken, NJ, USA, 1987; p. 101.

Review

Quantum Treatment of Inelastic Interactions for the Modeling of Nanowire Field-Effect Transistors

Youseung Lee [1], Demetrio Logoteta [2,†], Nicolas Cavassilas [2], Michel Lannoo [2], Mathieu Luisier [1] and Marc Bescond [3,*]

1 Integrated Systems Laboratory, ETH Zürich, 8092 Zürich, Switzerland; youseung.lee@iis.ee.ethz.ch (Y.L.); mluisier@iis.ee.ethz.ch (M.L.)
2 IM2NP, UMR CNRS 7334, Aix-Marseille Université, Technopôle de Château-Gombert, Bâtiment Néel, 60 Rue Frédéric Joliot Curie, 13453 Marseille, France; logotetad@gmail.com (D.L.); nicolas.cavassilas@im2np.fr (N.C.); michel.lannoo@free.fr (M.L.)
3 LIMMS, CNRS-UMI 2820, Institute of Industrial Science, University of Tokyo, Tokyo 153-8505, Japan
* Correspondence: bescond@iis.u-tokyo.ac.jp
† Current address: C2N-UMR 9001 CNRS/Université Paris-Sud-Université Paris-Saclay, 10 Boulevard Thomas Gobert, 91120 Palaiseau, France.

Received: 25 November 2019; Accepted: 16 December 2019; Published: 21 December 2019

Abstract: During the last decades, the Nonequilibrium Green's function (NEGF) formalism has been proposed to develop nano-scaled device-simulation tools since it is especially convenient to deal with open device systems on a quantum-mechanical base and allows the treatment of inelastic scattering. In particular, it is able to account for inelastic effects on the electronic and thermal current, originating from the interactions of electron–phonon and phonon–phonon, respectively. However, the treatment of inelastic mechanisms within the NEGF framework usually relies on a numerically expensive scheme, implementing the self-consistent Born approximation (SCBA). In this article, we review an alternative approach, the so-called Lowest Order Approximation (LOA), which is realized by a rescaling technique and coupled with Padé approximants, to efficiently model inelastic scattering in nanostructures. Its main advantage is to provide a numerically efficient and physically meaningful quantum treatment of scattering processes. This approach is successfully applied to the three-dimensional (3D) atomistic quantum transport OMEN code to study the impact of electron–phonon and anharmonic phonon–phonon scattering in nanowire field-effect transistors. A reduction of the computational time by about $\times 6$ for the electronic current and $\times 2$ for the thermal current calculation is obtained. We also review the possibility to apply the first-order Richardson extrapolation to the Padé $N/N-1$ sequence in order to accelerate the convergence of divergent LOA series. More in general, the reviewed approach shows the potentiality to significantly and systematically lighten the computational burden associated to the atomistic quantum simulations of dissipative transport in realistic 3D systems.

Keywords: quantum modeling; nonequilibrium Green's function; nanowire transistor; electron–phonon interaction; phonon–phonon interaction; self-consistent Born approximation; lowest order approximation; Padé approximants; Richardson extrapolation

---

1. Introduction

Recent advances in the nanostructure engineering have led to a vast variety of nano-scale material applications in different areas, e.g., electronics [1], photonics [2], and thermoelectric devices [3]. Among them, nanowire (NW) field-effect transistor (FET) represents the most promising architecture for the next generation logic switches capable to reach sub-10-nm gate lengths in mass production [4]. The versatility of this architecture resides mostly in the fact that it can provide an excellent electrostatic control thanks to the gate-all-around (GAA) configuration [5] and can incorporate novel materials exhibiting high mobilities and/or high band gap [6–11].

In order to move forward in the development of NW FETs, various quantum mechanical effects inside NWs should be theoretically investigated. In line with that purpose, there have been abundant theoretical achievements and discussions to shed light on quantum physics underlying electrical transport. Wigner function-based quantum transport approach [12–14], considered as a quantum counterpart of Boltzmann transport equation, is one of such achievements. Pauli master equation was also proposed to treat inelastic scattering processes by using the concept of scattering states based on Fermi's Golden rule [15,16]. Meanwhile, Bohmian mechanics has been applied to develop a many-body quantum transport simulator [17,18].

Along with all the methods aforementioned, the Nonequilibrium Green's function (NEGF) formalism, first developed by Gordon Baym, Leo P. Kadanoff [19,20], and Leonid V. Keldysh [21] in the 1960s, has attracted intensive interests due to its capability to address various quantum effects taking place in nanostructures [22–24]. In particular, during the last decade there have been significant attempts to apply NEGF formalism for describing quantum mechanical effects inside nano-scaled devices such as quantum confinement [25], tunneling [26,27], surface roughness scattering [28], and electron–phonon interactions [29]. Among those effects, the treatment of inelastic interactions within this formalism, based on the concept of scattering self-energy, has become widespread.

From a numerical point of view, however, a challenge still remains in treating self-energies, which is particularly evident in applying NEGF to three-dimensional (3D) realistic structures. The main bottleneck usually comes from the conventional treatment of the nonlinear Dyson's equation, within the self-consistent Born approximation (SCBA). Indeed, the SCBA satisfies the current conservation law via a Φ-derivable self-energy, that is, $\Sigma[G] = \delta\Phi/\delta G$, but it requires a huge number of iterations to meet this condition. This is due, in part, to the innate characteristics of the SCBA algorithm, that works by including in the approximate solution higher-order Feynman diagrams at each iteration, regardless of if they are or not current conserving. The SCBA iteration process stops when the conservation law is satisfied, that is, when conserving diagrams are dominant over non-conserving ones. As a consequence, the implementation of the SCBA algorithm in atomistic 3D NEGF codes is usually only manageable with the help of supercomputer environments comprising several hundreds of CPUs/GPUs.

In this article, we review a highly efficient method [30–34], the so-called lowest order approximation (LOA) analytically continued by Padé approximants, to treat inelastic interactions within the NEGF framework. The main idea behind the method is to collect only the low-order scattering diagrams that guarantee the conservation of the current. Several analytic continuation techniques (e.g., Shanks transformation [35], Borel-Padé resummation [36], and hyper geometric resummation technique [37]) can be applied to the LOA series in order to reconstruct physical observables when the series is divergent. Here, we only focus on a rescaling technique [33,34] in combination with the Padé approximants [36,38], since it is very simple to implement in conventional NEGF codes. In the case of strongly divergent LOA series, we also show the application of the first-order Richardon extrapolation to the Padé $N/N-1$ approximants. The main advantage of this approach is that it avoids a large number of iterations while obtaining a relatively high degree of accuracy with respect to SCBA results.

To benchmark our method against the conventional SCBA scheme, we show its efficiency and accuracy in the calculation of electronic (including electron–phonon scattering) and phonon thermal (including anharmonic phonon–phonon scattering) currents flowing through a GAA NW-FET. Our investigation focuses on an n-type square cross-sectional silicon (Si) GAA NW-FET crystallographically oriented in $\langle 100 \rangle$ transport direction. As key findings, we show that the third-order LOA electronic currents analytically continued by Padé approximants can reproduce SCBA results within an error $\leqslant$10%. A similar accuracy is found by applying the first-order Richardson extrapolation to the sequence of Padé approximants of the thermal current.

The rest of the paper is organized as follows: Section 2 describes the theory of the LOA analytically continued by Padé approximants and compares it to the conventional SCBA iterative scheme. A simple rescaling technique to calculate the LOA series directly from the SCBA algorithm will also be explained. In Section 3, we will show that the approach can be advantageously applied to the computation of the electron and phonon currents of GAA NW-FETs. Finally, Section 4 will conclude the article with the key findings and the outlook of the method.

2. General Theoretical Framework

In this section, we theoretically detail the LOA approach for the treatment of inelastic interactions within the NEGF framework [29,39–41] and we compare it to the conventional SCBA. We then present a simple rescaling technique, computationally most efficient and applicable to 3D realistic structures, to calculate the LOA expectation values from the conventional SCBA method. A matrix form of Padé approximants and the first-order Richardson extrapolation technique are also discussed.

2.1. Dyson Equation

In the NEGF theory, the Dyson equation links the noninteracting Green's function g_0 to the fully interacting Green's function G *via* the scattering self-energy Σ, which describes all the interactions. In a simplified matrix notation, we have

$$G = g_0 + g_0\Sigma[G]G, \tag{1}$$

with the abbreviations i = (r_i, t_i), $g_0\Sigma[G]G = \int d2 \int d2' g_0(1;2)\Sigma(2;2')G(2';1')$. The Feynman diagrams corresponding to electron–phonon scattering are shown in Figure 1. Since the scattering self-energy is a functional of G, i.e., $\Sigma = \Sigma[G]$, the Dyson equation is nonlinear and suitable approximations are required for the self-energy. The Φ-derivable approximation in the Luttinger–Ward picture [42,43] is a prescription to construct a self-energy satisfying the current conservation law. Proper resolutions of the Dyson equation should then be handled with a Φ-derivable self-energy. While both the conventional SCBA and the LOA provide a Φ-derivable self-energy, they solve the Dyson equation by using two different kind of algorithms, iterative and direct, respectively.

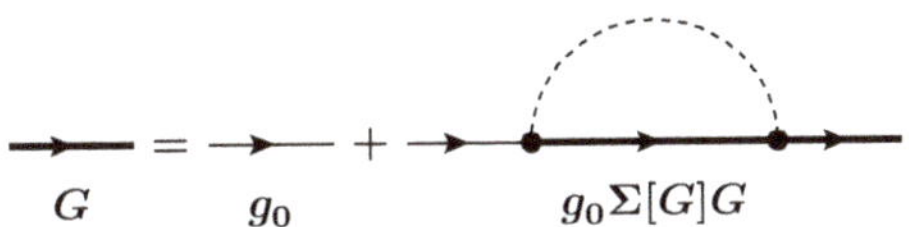

Figure 1. Feynman diagrams for the Dyson equation showing the relation between the noninteracting electron Green's function g_0 (thin lines with an arrow) and fully interacting one G (thick lines with an arrow): The dashed line denotes the free phonon propagator.

2.2. Self-Consistent Born Approximation

The nonlinearity of the Dyson equation leads itself naturally to an iterative scheme to obtain solutions. This inspiration then led to the application of the Born approximation in the concept of self-consistency, i.e., SCBA, which is commonly considered essential for Φ-derivability. By assuming $G_N \simeq G_{N-1}$ with very large N, Equation (1) can be rewritten as

$$G_N = [g_0^{-1} - \Sigma[G_{N-1}]]^{-1}, \tag{2}$$

or in a Taylor series expansion form

$$G_N = g_0 + g_0\Sigma[G_{N-1}]g_0 + g_0\Sigma[G_{N-1}]g_0\Sigma[G_{N-1}]g_0 + \cdots . \tag{3}$$

where G_N stands for the Green's function at the Nth iteration step. The SCBA Green's functions G_1 and G_2 are then defined from Equation (3) as

$$G_1 = g_0 + g_0\Sigma[g_0]g_0 + g_0\Sigma[g_0]g_0\Sigma[g_0]g_0 + \cdots . \tag{4}$$

and

$$\begin{aligned} G_2 &= g_0 + g_0\Sigma[G_1]g_0 + g_0\Sigma[G_1]g_0\Sigma[G_1]g_0 + \cdots , \\ &= g_0 + g_0\Sigma[g_0]g_0 + (g_0\Sigma[g_0]g_0\Sigma[g_0]g_0 + g_0\Sigma\left[g_0\Sigma\left[g_0\right]g_0\right]g_0) \\ &\quad + g_0\Sigma[g_0]g_0\Sigma[g_0]g_0\Sigma[g_0]g_0 + \cdots , \end{aligned} \tag{5}$$

respectively. As shown in Figure 2a,b, each SCBA Green's function includes an infinite number of diagrams. Therefore, G_1 and G_2 do not necessarily preserve the conservation law since there are non-conserving higher-order diagrams, according to the corresponding scattering order [31]. This character of the SCBA scheme allows only an asymptotic approach to the conservation law, that is, a large number of iterations with specified convergence criteria depending on the scattering strength of the system.

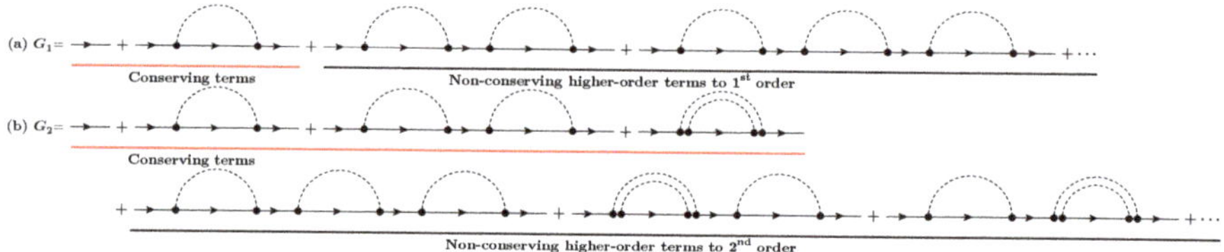

Figure 2. Feynman diagrams for self-consistent Born approximation (SCBA) Green's functions, (**a**) G_1 (at first iteration) and (**b**) G_2 (at second iteration): Conserving (red-underlined) and non-conserving (black-underlined) terms are arranged according to ascending interaction order. Noninteracting electron Green's functions are described by thin lines with an arrow while free phonon propagators are represented by dashed lines.

2.3. Lowest Order Approximation

The fact that the SCBA Green's functions include non-conserving diagrams has inspired the concept of the LOA method, i.e., calculating only the conserving diagrams based on the Φ-derivability at each scattering order, as shown in Figure 3a,b. The first-order LOA in the case of electron–phonon scattering has been suggested in Reference [30] by treating the term $g_0\Sigma[G]G$ from Equation (1) as a perturbation δG and by applying a Taylor series expansion to the corresponding self-energy to obtain the first-order LOA Green's function g_{1LOA} as

$$g_{1LOA} = g_0 + g_0\Sigma[g_0]g_0, \tag{6}$$

where the interaction self-energy is constructed only by the noninteracting Green's function g_0 and is Φ-derivable since $\Sigma[g_0] = \delta\Phi[g_0]/\delta g_0$. The work in Reference [30] showed that, in the weak-scattering regime, the expectation value constructed from the first-order LOA Green's function (therein, 1st LOA current) is very similar to the one from the SCBA. However, it was also shown that the spectral currents built by the 1st LOA Green's function are very far from the SCBA results since only the first order in the interaction was included.

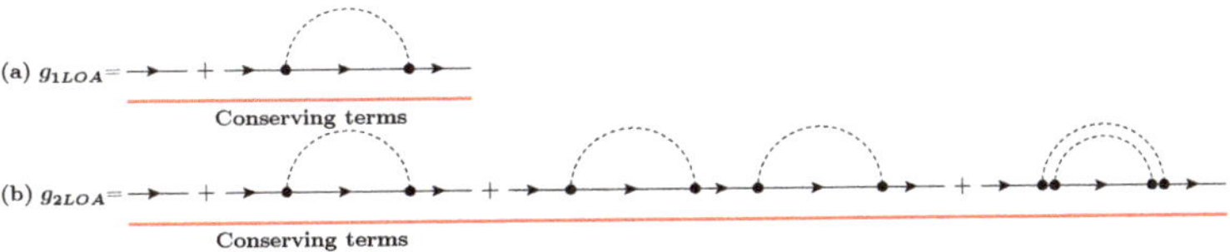

Figure 3. Feynman diagrams for Lowest Order Approximation (LOA) Green's functions: (**a**) g_{1LOA} at the first order and (**b**) g_{2LOA} at the second order in interactions. Conserving terms are red-underlined. Noninteracting electron Green's functions are described by thin lines with an arrow while free phonon propagators are represented by dashed lines.

In References [31,32], it was shown that, by using the term $\Delta g_1 = g_0\Sigma[g_0]g_0$ from Equation (6) as a basic building block, the generalized LOA expansion series can be built as

$$g_N = g_0 + \sum_{n=1}^{N} \Delta g_n, \tag{7}$$

where $g_N(N > 1)$ is the LOA Green's function at Nth-order in the interaction and Δg_n is the perturbation term of order n in the interaction. Since the Dyson equation has a recursion relation $g_0\Sigma[G]G$, the general expression of g_N can be obtained by injecting Equation (7) into Equation (1) such that

$$g_0 + g_0\Sigma[g_{N-1}]g_{N-1} = g_0 + g_0\Sigma[g_0 + \sum_{n=1}^{N-1} \Delta g_n]\,(g_0 + \sum_{n=1}^{N-1} \Delta g_n), \tag{8}$$

and then by discarding all higher-order terms to include only N-order interactions for g_N. For example, by following the previous process, g_2 and g_3 are constructed as

$$g_2 = g_1 + g_0\Sigma_1\Delta g_1 + g_0\Sigma_2\Delta g_0 = g_1 + \Delta g_2, \tag{9}$$

$$g_3 = g_2 + g_0\Sigma_1\Delta g_2 + g_0\Sigma_2\Delta g_1 + g_0\Sigma_3\Delta g_0 = g_2 + \Delta g_3, \tag{10}$$

where the corresponding self-energy is defined such that Σ_n depends on the $n-1$ order perturbation term

$$\Sigma_n = \Sigma\left[\Delta g_{n-1}\right]. \tag{11}$$

As a result, the perturbation term for $N \geqslant 1$ is as follows:

$$\Delta g_N = g_0 \sum_{n=0}^{N-1} \Sigma_{N-n}\Delta g_n, \tag{12}$$

with

$$\Delta g_1 = g_0\Sigma[g_0]g_0, \tag{13}$$

and

$$g_N = g_{N-1} + \Delta g_N. \tag{14}$$

The overall schematic of practical calculations for the LOA Green's function g_N, which was explained in Reference [32] in detail, is summarized in Algorithm 1 below.

Algorithm 1 Nth-order LOA calculation.

$N \leftarrow$ <the order of LOA>
for $i \leftarrow 0$ *to* N **do**
  if $i = 0$ **then**
    $g_0^r \leftarrow [EI - H - \Sigma_C^r]$, E (Energy), H (Hamiltonian), and Σ_C (Contact self-energy)
    $g_0^{\lessgtr} \leftarrow g_0^r \Sigma_C^{\lessgtr} g_0^a$, the superscript r (Retarded), a (Advanced), and $\lessgtr$ (Lesser/Greater)
    $\Delta g_0^{\lessgtr} \leftarrow g_0^{\lessgtr}, \Delta g_0^r \leftarrow g_0^r$
  else
    for $n \leftarrow 0$ *to* $i - 1$ **do**
      : interacting self energy calculation
      $\Sigma_{int,i}^{\lessgtr}$ and $\Sigma_{int,i}^r$
      : perturbation term calculation
      $temp(\Delta g_i^r) \leftarrow g_0^r \Sigma_{int,i-n}^r \Delta g_n^r$
      $\Downarrow$ applying Langreth Theorem
      $temp(\Delta g_i^{\lessgtr}) \leftarrow g_0^r \Sigma_{int,i-n}^r \Delta g_n^{\lessgtr} + g_0^r \Sigma_{int,i-n}^{\lessgtr} \Delta g_n^a + g_0^{\lessgtr} \Sigma_{int,i-n}^a \Delta g_n^a$
      $\Delta g_i^r \leftarrow \Delta g_i^r + temp(\Delta g_i^r)$
      $\Delta g_i^{\lessgtr} \leftarrow \Delta g_i^{\lessgtr} + temp(\Delta g_i^{\lessgtr})$
    end for
    $g_i^r \leftarrow g_{i-1}^r + \Delta g_i^r$
    $g_i^{\lessgtr} \leftarrow g_{i-1}^{\lessgtr} + \Delta g_i^{\lessgtr}$
  end if
end for

Equations (12)–(14) show that, by construction, each LOA Green's function g_N includes only N-order interactions, meaning that it can generate the corresponding expectation value $\mathcal{O}_N = \mathcal{O}(g_N)$. We then obtain the LOA series for any expectation value. For example, in this article, electronic current series (including electron–phonon scattering) and thermal current series (including phonon–phonon scattering) are defined as $\mathcal{I}_N = \mathcal{I}(g_N)$ and $\mathcal{Q}_N = \mathcal{Q}(g_N)$, respectively, to any order in the interaction:

$$\mathcal{I}_N = \mathcal{I}_0 + \sum_{n=1}^{N} \Delta\mathcal{I}_n, \; \mathcal{Q}_N = \mathcal{Q}_0 + \sum_{n=1}^{N} \Delta\mathcal{Q}_n, \tag{15}$$

where $\Delta\mathcal{I}_n = \mathcal{I}_n - \mathcal{I}_{n-1}$ and $\Delta\mathcal{Q}_n = \mathcal{Q}_n - \mathcal{Q}_{n-1}$ represent the difference between the expectation values of the nth and $(n-1)$th orders and where $\mathcal{I}_0$ and $\mathcal{Q}_0$ are the expectation values from the noninteracting Green's function g_0 (here, the ballistic one).

Indeed, with Algorithm 1, one can compute the exact LOA Green's function at any order within the considered interaction. However, the Langreth theorem, involved in transforming the contour integration into real-time integration, leads to a large number of matrix inversions and multiplications for higher-order LOA calculations, which restricts the applicability of the method. Therefore, in the next section, we present a simple rescaling technique to directly calculate the LOA expectation values from the SCBA results.

2.4. Rescaling Technique

The rescaling technique has been developed in Reference [33] to overcome the limitation of Algorithm 1 in the spirit of preserving the conventional SCBA algorithm since the latter algorithm is computationally cheap as long as a small number of iterations is needed. The method was applied to the case of electron–phonon scattering inside GAA NW-FETs, described by using an $sp^3d^5s^*$ tight-binding (TB) model, as implemented in the atomistic code OMEN [29,44]. Later on, Lee et al. [34] have shown that the method is also applicable to describe anharmonic phonon–phonon scattering in which the anharmonic self-energy Π is a function of the square of phonon Green's functions D.

The basic idea of the rescaling technique is that, by multiplying a properly chosen factor $1/\lambda$ to any scattering self-energy (e.g., $\Sigma[G]/\lambda$ for electron–phonon and $\Pi[DD]/\lambda$ for phonon–phonon), infinite non-conserving diagrams in SCBA Green's functions can be eliminated. The reason behind this is based on the fact that conserving and non-conserving diagrams in the SCBA are arranged according to ascending order in interactions, as shown in Figure 2.

For example, let us rewrite Equation (3), which is a Taylor expansion series of the SCBA Green's function, in a general form applicable to any Green's function ($Z = G$ for electron and $Z = D$ for phonon), as

$$\begin{aligned} Z_N = z_0 &+ z_0\Phi[(Z_{N-1})^l]z_0 \\ &+ z_0\Phi[(Z_{N-1})^l]z_0\Phi[(Z_{N-1})^l]z_0 + \cdots, \; l \geqslant 1, \end{aligned} \tag{16}$$

where z_0 is the noninteracting Green's function and l denotes the dependence of the self-energy on the Green's function to the lth power ($\Phi[Z_{N-1}] = \Sigma[G_{N-1}]$ with $l = 1$ for electron–phonon and $\Phi[Z_{N-1}Z_{N-1}] = \Pi[D_{N-1}D_{N-1}]$ with $l = 2$ for phonon–phonon). By introducing a scaling parameter λ_1 into the scattering self-energy and by proceeding to the first SCBA iteration, higher-order non-conserving diagrams vanish [33] since each term is scaled by $(1/\lambda_1)^n$ in the interaction order n. We then express the first-order LOA Green's function rescaled by λ_1 as

$$Z_1^{\lambda_1} = z_0 + \frac{1}{\lambda_1}\Delta z_1 \, (= z_0\Phi_1 z_0), \; \Phi_1 = \Phi\left[(z_0)^l\right], \tag{17}$$

where the nth-order ($n > 1$) non-conserving terms are suppressed thanks to the factors $(1/\lambda_1)^n$. The first-order LOA expectation value ($\mathcal{O}_1$) is then calculated by using λ_1 such as $\mathcal{O}_1 = \mathcal{O}_0 + \Delta\mathcal{O}_1 = \mathcal{O}_0(z_0) + \mathcal{O}(z_0\Phi[(z_0)^l]z_0) = \mathcal{O}_0(z_0) + \lambda_1[\mathcal{O}(Z_1^{\lambda_1}) - \mathcal{O}(z_0)]$.

For the second-order calculation, let us assume that we have the first-order LOA Green's function rescaled by λ_2 as $Z_1^{\lambda_2} = z_0 + \frac{1}{\lambda_2}\Delta z_1$. The second SCBA iteration with a properly chosen scaling factor λ_2 then produces the self-energy Φ_2 as

$$\begin{aligned} \Phi_2 = \Phi\left[\frac{(Z_1^{\lambda_2})^l}{\lambda_2}\right] &= \Phi\left[\frac{1}{\lambda_2}(z_0 + \frac{1}{\lambda_2}\Delta z_1)^l\right], \\ =&\Phi\left[\frac{(z_0)^l}{\lambda_2} + \frac{z_0^{l-1}\Delta z_1}{\lambda_2^2} + \cdots + \frac{z_0\Delta z_1^{l-1}}{\lambda_2^l} + \frac{(\Delta z_1)^l}{\lambda_2^{l+1}}\right]. \end{aligned} \tag{18}$$

In Equation (18), only the terms related to $(1/\lambda_2)$ and $(1/\lambda_2)^2$ are retained while the other terms are omitted. Therefore, the second-order LOA Green's function $Z_2^{\lambda_2}$ rescaled by λ_2 also does not include all the terms related to $(1/\lambda_2)^N$ ($N > 2$). Reconstructing the second-order LOA expectation value then follows the way explained in Reference [33]. This process can be generalized to any order LOA Green's function. Given an exact Green's function at the $(N-1)$th order with the rescaling factor λ_N as

$$Z_{N-1}^{\lambda_N} = z_0 + \frac{1}{\lambda_N}\Delta z_1 + \cdots + \frac{1}{\lambda_N^{N-1}}\Delta z_{N-1}, \tag{19}$$

an exact self-energy at the Nth-order can be obtained since the scaling factor λ_N automatically cancels out all the higher-order interactions (> Nth-order): if Z_{N-1} is correct to the $(N-1)$th order, $\Phi_N\left[(Z_{N-1})^l\right]$ is also correct to the Nth-order [34]. By using the proposed method, any-order LOA expectation values can then be easily obtained from the SCBA iterations.

2.5. Matrix Form of the Padé Approximants

The LOA Green's function, computed either by the direct algorithm or by the rescaling technique, consists of finite conserving diagrams up to the Nth-order, i.e., forms a truncated perturbation-expansion series. This series depends on an interaction parameter U, always associated to two vertices in the electron–phonon and phonon–phonon scattering diagrams:

$$z_0 + \Delta z_1(U) + \cdots + \Delta z_N(U^N). \tag{20}$$

As in the general case of perturbation series, the LOA series has a radius of convergence U_r, then being convergent ($U < U_r$) or divergent ($U > U_r$), depending on U [36,38,45]. Therefore, in order to obtain meaningful values for the physical observables, the application of resummation techniques (e.g., Padé approximants [31–34,46,47], hypergeometric resummation [37], and Padé + Richardson extrapolation [34]) is generally required. Here, we review a matrix form of Padé approximants [48] since it has been already proved to be very efficient with a relatively high accuracy in the previous works [32–34].

The Padé approximant transforms a power series function $F_N(x)$ into a fractional function including two polynomials, $P_p(x)$ as the numerator and $M_m(x)$ as the denominator [49] such as

$$F_N(x) = \sum_{k=0}^{p+m} f_k x^k = f_0 + f_1 x + f_2 x^2 + \cdots + f_{p+m} x^{p+m} = \frac{p_0 + p_1 x + p_2 x^2 + \cdots + p_p x^p}{1 + m_1 x + m_2 x^2 + \cdots + m_m x^m} = F_{p/m}(x). \tag{21}$$

By matching coefficients at the same order between two functions, $l + m + 1$ linear homogeneous equations can be obtained as

$$\begin{aligned} p_0 &= f_0, \\ p_1 - f_0 m_1 &= f_1, \\ p_2 - f_0 m_2 - f_1 m_1 &= f_2, \\ &\vdots \\ p_p - f_0 m_p - f_1 m_{p-1} - \cdots - f_{p-1} m_1 &= f_p, \\ -f_{p-m+1} m_m - f_{p-m+2} m_{m-1} - \cdots - f_p m_1 &= f_{p+1}, \\ -f_{p-m+2} m_m - f_{p-m+3} m_{m-1} - \cdots - f_{p+1} m_1 &= f_{p+2}, \\ &\vdots \\ -f_p m_m - f_{p+1} m_{m-1} - \cdots - f_{p+m-1} m_1 &= f_{p+m}. \end{aligned} \tag{22}$$

We then obtain a matrix form $\mathbb{A}\boldsymbol{x} = \boldsymbol{b}$ to calculate Padé coefficients $p_0, \ldots, p_p$ and $m_1, \ldots, m_m$ as

$$\begin{bmatrix} 1 & 0 & \cdots & 0 & 0 & 0 & \cdots & 0 \\ 0 & 1 & \cdots & 0 & -f_0 & 0 & \cdots & 0 \\ \vdots & & \ddots & \vdots & \vdots & & \ddots & \vdots \\ 0 & 0 & \cdots & 1 & -f_{p-1} & -f_{p-2} & \cdots & 0 \\ 0 & 0 & \cdots & 0 & -f_p & -f_{p-1} & \cdots & -f_{p-m+1} \\ 0 & 0 & \cdots & 0 & -f_{p+1} & -f_p & \cdots & -f_{p-m+2} \\ \vdots & & \ddots & \vdots & \vdots & & \ddots & \vdots \\ 0 & 0 & \cdots & 0 & -f_{p+m-1} & -f_{p+m-2} & \cdots & -f_p \end{bmatrix} \begin{bmatrix} p_0 \\ p_1 \\ \vdots \\ p_p \\ m_1 \\ m_2 \\ \vdots \\ m_m \end{bmatrix} = \begin{bmatrix} f_0 \\ f_1 \\ \vdots \\ f_p \\ f_{p+1} \\ f_{p+2} \\ \vdots \\ f_{p+m} \end{bmatrix}. \tag{23}$$

By solving Equation (23), a power series-transformed fractional function can be determined as $F_{p/m}(x) = \frac{p_0 + p_1 x + p_2 x^2 + \cdots + p_p x^p}{1 + m_1 x + m_2 x^2 + \cdots + m_m x^m}$. By appling this technique to the Nth-order LOA expectation values $\mathcal{O}_N$, Padé analytically continued expectation values $\mathcal{O}_{p/m}$ with $N = p + m$ are obtained.

2.6. Richardson Extrapolation

In this subsection, we introduce the first-order Richardson extrapolation technique, applicable to accelerate the convergence of divergent LOA expectation values analytically continued by Padé $N-1/N$ approximants toward SCBA values [34].

Let us assume that the LOA + Padé sequence is a monotone series S_N approaching the SCBA result S. We then write S_N in an asymptotic form as [50]

$$S_N \approx S + \frac{c_1}{N} + \frac{c_2}{N^2} + \frac{c_3}{N^3} + \cdots, \tag{24}$$

with unknown coefficients $c_1, c_2, \cdots$. By considering the first-order approximation of S based on two consecutive terms S_N and S_{N+1},

$$N(S_N - S) = c_1, \quad (N+1)(S_{N+1} - S) = c_1, \tag{25}$$

the first-order Richardson extrapolation can be obtained as

$$S_N^{[1]} = (N+1)S_{N+1} - NS_N. \tag{26}$$

This technique can be applied to the sequence of LOA expectation values analytically continued by the Padé $N-1/N$ approximants ($\mathcal{O}_{N-1/N}$). For example, the expectation values obtained from Padé $0/1(\mathcal{O}_{0/1})$ and Padé $1/2$ ($\mathcal{O}_{1/2}$) become the first and second elements of the sequence, i.e., $S_1 = \mathcal{O}_{0/1}$ and $S_2 = \mathcal{O}_{1/2}$, respectively (in general, $S_N = \mathcal{O}_{N-1/N}$). By using Equation (26) with $\mathcal{O}_{0/1}$ and $\mathcal{O}_{1/2}$, we can then predict SCBA values as

$$S_1^{[1]} = 2S_2 - S_1 = 2\mathcal{O}_{1/2} - \mathcal{O}_{0/1} \approx \mathcal{O}_{SCBA}. \tag{27}$$

3. Applications to Electron and Phonon Transports in a Nanowire Transistor

In this section, we benchmark the performance of the LOA + Padé approach against the SCBA scheme in the description of electron–phonon and anharmonic phonon–phonon scattering inside a 3D nano-device. The investigated device is a Si GAA NW-FET with a 3 nm $\times$ 3 nm square cross section where electron or phonon transport occurs along the $\langle 100 \rangle$ crystallographic direction (Figure 4). In such a small cross-sectional NW, the transport is dominated by quantum confinement effects. The corresponding electronic band-structure and phonon dispersion relation are shown in Figure 5a,b, respectively.

The coupling effects between electronic and anharmonic thermal currents are not considered since the aim of the benchmarking is only to test the efficiency and accuracy of the method for each scattering mechanism. The electronic and thermal currents are then separately calculated in different configurations as follows. (*i*) For electron transport, the gate length is $L_G = 13$ nm while the source and drain extension lengths are $L_{S/D} = 10$ nm, with a doping concentration of donors $N_{S/D} = 1 \times 10^{20}$ cm^{-3}. A 1-nm-thick silicon dioxide layer surrounds the NW structure. (*ii*) In the case of phonon transport, an ungated 60-nm-long NW structure without oxide layers is considered. All atoms on the NW surface are free to move. For the sake of clarity, only undoped NWs are considered. Under this condition, a very small number of electrons participate in the thermal conduction. This allows us to safely neglect the impact of eletron–phonon scattering [51].

In order to validate the LOA + Padé approach, we use the 3D atomistic NEGF code OMEN [29,44,51–53] in which a $sp^3d^5s^*$ TB model for electrons and a modified valence-force-field (VFF) method for phonons are implemented. OMEN is among the most sophisticated atomistic simulators of quantum transport of nano-devices. Indeed, the latest version of OMEN also includes density-functional theory (DFT)-based Hamiltonian expressed in a maximally localized Wannier function basis [10]. However, it requires considerable computational resources, especially when the

SCBA loop is brought into play. In the following Sections 3.1 and 3.2, the LOA approach within the rescaling technique formulation is used to calculate the electronic and thermal current. These results are benchmarked against the values obtained by running the full SCBA loop in OMEN until the current is conserved within a tolerance of 1%.

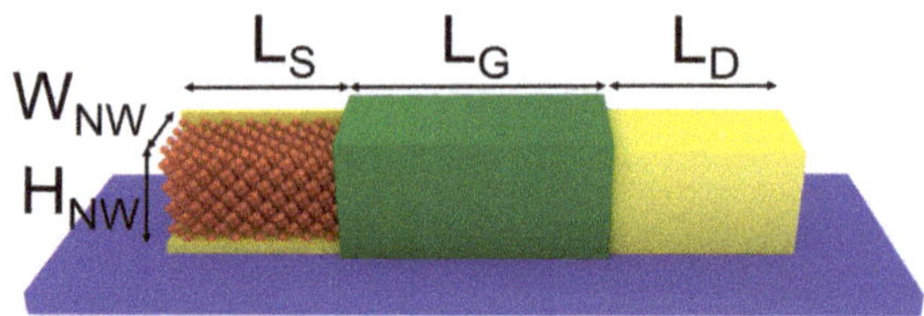

Figure 4. Schematic view of a Si gate-all-around (GAA) nanowire (NW) field-effect transistor (FET) crystallographically oriented along the $\langle 100 \rangle$ direction with a $H_{NW}(3\,\mathrm{nm}) \times W_{NW}(3\,\mathrm{nm})$ square cross section: (Electron transport) The gate length is $L_G = 13$ nm, and the length of source/drain region measures $L_{S/D} = 10$ nm. The NW structure is surrounded by a 1-nm-thick silicon dioxide layer. The concentration of donors in the source and drain regions is $N_{S/D} = 1 \times 10^{20}$ cm^{-3}. (Phonon transport) The NW total length ($L_G + L_S + L_D$) is 60 nm. The NW is undoped, ungated, and free of the oxide layer.

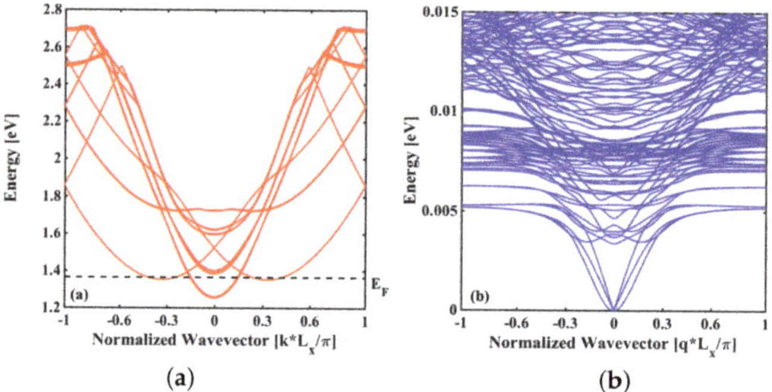

(**a**) (**b**)

Figure 5. (**a**) Electronic conduction band-structure of the Si NW sketched in Figure 4, obtained with a full band tight-binding $sp^3d^5s^*$ model without spin-orbit coupling. (**b**) Phonon dispersion relation obtained with a modified valence-force-field method. L_x is a slab length.

3.1. Electron–Phonon Scattering in a Nanowire Transistor

Here, we investigate steady-state electron transport in which electrons are coupled with the equilibrium phonon bath at room temperature. A full band $sp^3d^5s^*$ TB model without spin-orbit coupling [54,55] is employed to describe the electronic states while a modified VFF method including four bond-interaction terms (harmonic approximations) [56,57] is used for the description of the NW phonon bath. The diagonal lesser self-energy for electron-phonon scattering is

$$\Sigma^<_{nn}(E) = \mathrm{i}\sum_{l}\sum_{\lambda,\mathbf{q}}\int_{-\infty}^{\infty}\frac{\mathrm{d}(E')}{2\pi}\mathbf{M}^{\lambda}_{nl}(\mathbf{q})d^<_{0,\lambda}(\mathbf{q},E-E')G^<_{ll}(E')\mathbf{M}^{\lambda*}_{ln}(\mathbf{q}), \tag{28}$$

with

$$\mathbf{M}^{\lambda}_{nl}(\mathbf{q}) = \sqrt{\frac{\hbar}{2\omega_\lambda(\mathbf{q})}}\sum_i \nabla_i H_{nl}\left[\frac{f^i_\lambda(\mathbf{R}_l,\mathbf{q})}{\sqrt{m_l}} - \frac{f^i_\lambda(\mathbf{R}_n,\mathbf{q})}{\sqrt{m_n}}\right], \tag{29}$$

and

$$d^<_{0,\lambda}(\mathbf{q},\hbar\omega) = -2\pi\mathrm{i}\left[\langle n_\lambda(\mathbf{q})\rangle\,\delta(\hbar\omega \mp \hbar\omega_\lambda(\mathbf{q})) + (\langle n_\lambda(\mathbf{q})\rangle + 1)\delta(\hbar\omega \pm \hbar\omega_\lambda(\mathbf{q}))\right], \tag{30}$$

where $\hbar$ is the reduced Planck constant, E is the electron energy, $\mathbf{R}$ is the position vector, m is the atomic mass, f is the phonon displacement, and ω is the phonon frequency. The indices n and l denote the atomic positions, while λ and $\mathbf{q}$ represent the phonon mode λ with momentum $\mathbf{q}$. $\langle n_\lambda(\mathbf{q})\rangle$ is the expectation value of the phonon occupation in the mode λ and momentum $\mathbf{q}$ in thermal equilibrium. In the scattering self-energy Equation (28), the lesser Green's function $G^<$ is coupled *via* the nearest-neighbor matrix elements $\mathbf{M}$ to the lesser phonon Green's function $d_0^<$. The coupling matrix elements are obtained from the first derivative of the TB Hamiltonian matrix $\nabla_i H_{nl}$ between atoms n and l along the ith direction (x, y, or z). The real part of the self-energy, as commonly assumed in quantum transport nano-device modeling [58], is neglected.

By introducing three different scaling factors λ_1, λ_2, and λ_3 into Equation (28), as explained in Section 2.4, the LOA expectation values for electronic currents up to the third order can be obtained over a wide range of gate biases. The Padé approximant is then applied to analytically continue the LOA results. Figure 6 shows the drain current (I_D) vs. gate voltage (V_G) transfer characteristics obtained in the ballistic regime and by the SCBA, the 1st LOA, and the LOA + Padé 0/1 and 1/2 technique. It is clearly shown that 1st- and 3rd-order LOA currents combined with Padé 0/1 and 1/2, respectively, provide a fair evaluation of the SCBA result over the whole applied bias. The 1st LOA current is also very similar to the SCBA current except at high bias (i.e., $V_G = 0.6$ V), due to the particularly strong electron–phonon scattering in this regime [29]. For this strong scattering regime, the Richardson extrapolation has been applied, thereby obtaining an accelerated convergence of the strongly divergent LOA series. More precisely, the accuracies at $V_G = 0.6$ V are 89%, 93%, and 98% for the Padé 0/1, Padé 1/2, and Richardson, respectively. The values of the current at $V_G = 0.6$ V obtained within the different approximations are reported in Table 1, along with the required computational burden in terms of number of SCBA iterations. These values show that the LOA approach combined with the Padé analytical continuation and the Richardson extrapolation is able to approximate the SCBA results within an error of 2% in 6 SCBA iterations, while the full SCBA loop needs 35 iterations to converge.

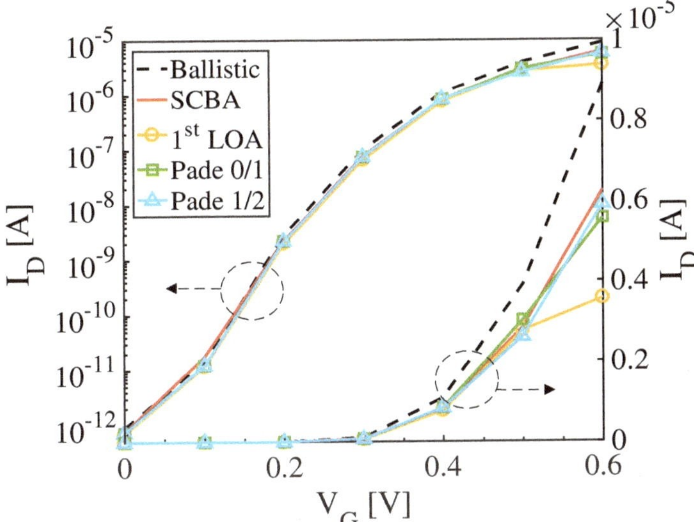

Figure 6. $I_D - V_G$ transfer characteristics of the n-type 3 nm × 3 nm square cross-sectional Si GAA NW-FET obtained for ballistic regime and by SCBA, 1st-order LOA, Padé 0/1, and Padé 1/2.

In a previous work, this approach has been applied to n- and p-type circular Si and Ge GAA NW FETs along the three main crystallographic orientations (<100>,<110>,<111>). A general similar conclusion about the efficiency of the LOA + Padé 1/2 has been reported, with a warning signal concerning the accuracy of the present approach for the situation where the 1st LOA current is larger than the ballistic one. Please refer to Reference [33] for the details.

Table 1. Accuracy ($\varepsilon = 100 \times |\mathcal{I}_{SCBA} - \mathcal{I}|/\mathcal{I}_{SCBA}$ where $\mathcal{I}$ is ballistic, LOA, Padé, or Richardson current) and efficiency (# of iterations) comparisons of 1st and 3rd LOA, Padé 0/1 and 1/2, and the corresponding Richardson currents with ballistic and SCBA currents at $V_G = 0.6$ V.

	Ballistic	LOA1	LOA3	Padé 0/1	Padé 1/2	Richardson	SCBA
Current [A]	8.9×10^{-6}	3.57×10^{-6}	-2.616	5.6×10^{-6}	5.9×10^{-6}	6.1×10^{-6}	6.3×10^{-6}
ε [%]	42.1	42.0	4.2e7	10.8	6.4	2.0	0.0
Number of iterations	0	1	6	1	6	6	35

3.2. Anharmonic Phonon–Phonon Scattering in A Nanowire

In this subseciton, we apply the LOA method to the steady-state anharmonic phonon transport. As described in Reference [51], a modified VFF method with harmonic bond interactions is used to construct the dynamical matrix, by which phonon frequencies are calculated. The noninteracting Green's function (here, the ballistic Green's function) is then calculated from the dynamical matrix and corresponding phonon frequency with the open boundary self-energy. Anharmonic effects, usually attributed to the third- and fourth-order contributions of the quantized atomic displacement [22,59] are described by a perturbative approach, that is, by solving the Dyson equation with the scattering self-energy in which VFF anharmonic interactions are included.

In this work, we focus on a single second-order perturbation effect based on the third-order anharmonicity, which models the anharmonic decay of a high-energy phonon ($\omega + \omega'$) into two lower energy phonons (ω and ω') or vice versa [51,59]. According to the contribution of the corresponding order, the greater/lesser anharmonic phonon–phonon scattering self-energy can be written as follows [51]:

$$\Pi^{\gtrless S,l_2 l_3}_{\nu\sigma}(\omega) = 2i\hbar \sum_{l_1,l_1',l_1'',l_1'''} \sum_{\mu,\mu',\mu'',\mu'''} \int_{-\infty}^{\infty} \frac{d\omega'}{2\pi} dV^{(3)l_2 l_1 l_1'}_{\nu\mu\mu'} D^{\gtrless,l_1 l_1''}_{\mu\mu''}(\omega+\omega') \times D^{\lessgtr,l_1''' l_1'}_{\mu'''\mu'}(\omega') \, dV^{(3)l_1'' l_1''' l_3}_{\mu''\mu'''\sigma}, \tag{31}$$

with

$$dV^{(3)l_2 l_1 l_1'}_{\nu\mu\mu'} = \frac{\partial^3}{\partial \mathbf{R}^{l_2}_{\nu} \partial \mathbf{R}^{l_1}_{\mu} \partial \mathbf{R}^{l_1'}_{\mu'}} V_{anh}, \tag{32}$$

where the indices μ, ν, and σ designate atomic positions and $dV^{(3)l_2 l_1 l_1'}_{\nu\mu\mu'}$ is a simplified notation for the third derivative of the VFF anharmonic potential energy V_{anh} with respect to the νth atomic position in the l_2 direction (x, y, or z) to the μth atomic position in the l_1 direction and to the μ'^{th} atomic position in the l_1' direction. This term couples together the greater/lesser Green's functions $D^{\gtrless}(\omega+\omega')$ and $D^{\lessgtr}(\omega')$, associated to the high-energy and low-energy phonons, respectively.

Like in the previous case, we restrict ourselves to the diagonal approximation with the anharmonic-force parameters calibrated to experimental data [51] and neglect the real part of the scattering self-energy since it mainly contributes to an energy renormalization [39,40,60]. In order to calculate the third-order expectation values of the thermal currents, three rescaling factors are injected into Equation (31). Without temperature difference between the left and right ends of the NW, the structure of interest is in equilibrium at room temperature. We then apply a small temperature difference ($\Delta T = 0.1$ K) between both extremities of the NW to make the phonon thermal current flow.

First, we calculate the ballistic thermal current flowing through the NW structure in which the anharmonic phonon–phonon scattering is "turned off". Then, to include the anharmonic effects, we compute the thermal currents through the conventional SCBA scheme, LOA + Padé, and LOA + Padé + Richardson. Figure 7 shows a comparison between the three methods. It is shown that all the LOA + Padé currents satisfy the current conservation law since the non-conserving diagrams are removed by the rescaling technique. However, the LOA currents diverge far from the SCBA value (not shown), testifying the importance of the anharmonic scattering even at room temperature. The application of Padé 0/1, 1/1, and 1/2 approximants results in the convergent behavior of the results, with 65%, 80%, and 87% accuracy with respect to SCBA, respectively. By applying the first-order Richardson extrapolation to the LOA + Pade 0/1 and 1/2 values, we improve the result, obtaining an accuracy higher than 90% without any loss in the numerical efficiency, compared to the SCBA result. Particularly, the LOA + Padé 1/2 + Richardson approximation requires 6 SCBA iterations, while the full SCBA loop requires 13 iterations to achieve the convergence. More exhaustive investigations of the calculation of anharmonic thermal currents in circular Si and GeNWs at several temperatures are reported in Reference [34].

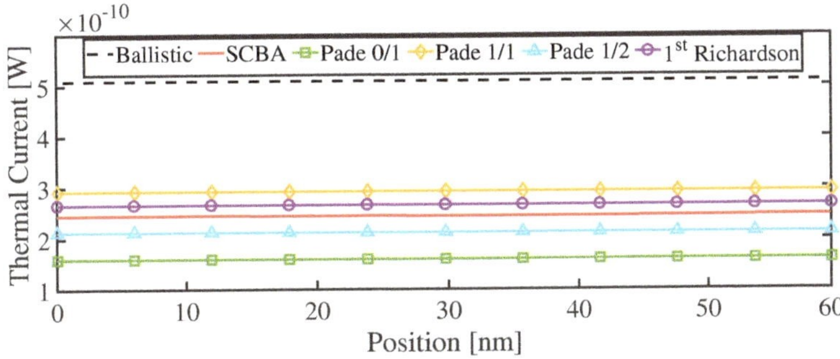

Figure 7. Thermal currents at room temperature in the 3 nm × 3 nm square cross-sectional Si NW of Figure 4 obtained for the ballistic regime and within the SCBA, Padé 0/1, Padé 1/1, Padé 1/2, and the first-order Richardson extrapolation.

4. Conclusions

In this work, we reviewed the theory, the so-called LOA approach combined with Padé approximants, developed within the NEGF formalism for the treatment of inelastic scattering. We benchmarked the method against the conventional SCBA approach by using an atomistic quantum transport code.

Differently from the SCBA, the LOA only includes current-conserving Feynman diagrams. This can results in a faster approximation of a current-conserving self-energy fulfilling the Dyson equation. We have demonstrated this case by investigating electron–phonon and anharmonic phonon–phonon scattering in a square Si GAA NW-FET. In the calculation of electronic and thermal currents, the LOA + Padé or the LOA + Padé + Richardson, implemented by a simple rescaling technique, successfully reproduced the SCBA values with a high accuracy (>90%) and computational efficiency (×6 faster for electronic currents and ×2 for thermal currents).

The main drawback of the approach is that the scaling factors used to calculate the higher-order perturbation terms (for both electron–phonon and phonon–phonon interactions) are determined via a rather pragmatic approach. Further work might be required to improve the choice of those scaling factors based on a more physically sound method.

Moreover, the method can be particularly advantageous (and thus worth to be investigated) in systems including localized states weakly coupled to the contacts, as for a deep quantum well. In these cases, the SCBA can take several thousands of iterations to converge and the use of LOA + Padé might be even more relevant.

Finally, it is worth noting that the reviewed method is also applicable to computationally more expensive codes such as a DFT-NEGF quantum transport simulator. In particular, it is expected that the LOA + Padé approach can provide more efficient results for device configurations where complex materials, e.g., transition-metal dichalcogenides, are involved, since the materials need an ab initio level of accuracy to describe their electronics properties and is, thus, costly. The approach should then faithfully broaden the accessibility of atomistic quantum transport codes based on the NEGF framework to investigate 3D realistic systems without the need for heavy computational resources.

Author Contributions: Y.L. conducted the numerical implementations. Y.L., M.L. (Mathieu Luisier), M.L. (Michel Lannoo) and M.B. developed the theoretical framework. M.L. (Mathieu Luisier) and M.B. supervised the work. All authors were involved in the interpretation and contributed to the final manuscript. All authors have read and agreed to the published version of the manuscript.

Funding: Y.L. acknowledges support from the Swiss National Science Foundation under the NCCR MARVEL. The project leading to this publication has received funding from Excellence Initiative of Aix-Marseille University—A*MIDEX, a French "Investissements d'Avenir" programme.

Conflicts of Interest: The authors declare no conflict of interest.

References

1. Suk, S.D.; Lee, S.Y.; Kim, S.M.; Yoon, E.J.; Kim, M.S.; Li, M.; Oh, C.W.; Yeo, K.H.; Kim, S.H.; Shin, D.S.; et al. High performance 5nm radius Twin Silicon Nanowire MOSFET (TSNWFET): Fabrication on bulk si wafer, characteristics, and reliability. In Proceedings of the IEEE International Electron Devices Meeting, Washington, DC, USA, 5 December 2005; pp. 717–720. [CrossRef]
2. Law, M.; Greene, L.E.; Johnson, J.C.; Saykally, R.; Yang, P. Nanowire dye-sensitized solar cells. *Nat. Mater.* **2005**, *4*, 455–459. [CrossRef] [PubMed]
3. Hochbaum, A.I.; Chen, R.; Delgado, R.D.; Liang, W.; Garnett, E.C.; Najarian, M.; Majumdar, A.; Yang, P. Enhanced thermoelectric performance of rough silicon nanowires. *Nature* **2008**, *451*, 163–167. [CrossRef] [PubMed]
4. Mertens, H.; Ritzenthaler, R.; Hikavyy, A.; Kim, M.S.; Tao, Z.; Wostyn, K.; Chew, S.A.; De Keersgieter, A.; Mannaert, G.; Rosseel, E.; et al. Gate-All-Around MOSFETs based on Vertically Stacked Horizontal Si Nanowires in a Replacement Metal Gate Process on Bulk Si Substrates. In Proceedings of the IEEE Symposium on VLSI Technology, Honolulu, HI, USA, 14–16 June 2016; pp. 1–2.
5. Capogreco, E.; Arimura, H.; Witters, L.; Vohra, A.; Porret, C.; Loo, R.; De Keersgieter, A.; Dupuy, E.; Marinov, D.; Hikavyy, A.; et al. High performance strained Germanium Gate All Around p-channel devices with excellent electrostatic control for sub-Jtlnm LG. In Proceedings of the Symposium on VLSI Technology, Kyoto, Japan, 9–14 June 2019; pp. T94–T95. [CrossRef]
6. Li, L.; Yu, Y.; Ye, G.J.; Ge, Q.; Ou, X.; Wu, H.; Feng, D.; Chen, X.H.; Zhang, Y. Black phosphorus field-effect transistors. *Nat. Nanotech.* **2014**, *9*, 372–377. [CrossRef] [PubMed]
7. Franklin, A.D.; Luisier, M.; Han, S.J.; Tulevski, G.; Breslin, C.M.; Gignac, L.; Lundstrom, M.S.; Haensch, W. Sub-10 nm Carbon Nanotube Transistor. *Nano Lett.* **2012**, *12*, 758. [CrossRef] [PubMed]
8. Radisavljevic, B.; Radenovic, A.; Brivio, J.; Giacometti, V.; Kis, A. Single-layer MoS2 transistors. *Nat. Nanotech.* **2011**, *6*, 147. [CrossRef] [PubMed]
9. Yoon, Y.; Ganapathi, K.; Salahuddin, S. How Good Can Monolayer MoS2 Transistors Be? *Nano Lett.* **2011**, *11*, 3768. [CrossRef]
10. Szabó, A.; Rhyner, R.; Luisier, M. Ab initio simulation of single- and few-layer MoS2 transistors: Effect of electron-phonon scattering. *Phys. Rev. B* **2015**, *92*, 035435. [CrossRef]
11. Cao, J.; Logoteta, D.; Özkaya, S.; Biel, B.; Cresti, A.; Pala, M.G.; Esseni, D. Operation and Design of van der Waals Tunnel Transistors: A 3-D Quantum Transport Study. *IEEE Trans. Electron Devices* **2016**, *63*, 1–7. [CrossRef]

12. Tatarskiĭ, V.I. The Wigner representation of quantum mechanics. *Sov. Phys. Uspekhi* **1983**, *26*, 311–327. [CrossRef]
13. Jacoboni, C.; Brunetti, R.; Bordone, P.; Bertoni, A. Quantum tansport and its simulation with the wigner-function approach. *Int. J. High Speed Electron. Syst.* **2001**, *11*, 387–423. [CrossRef]
14. Querlioz, D.; Saint-Martin, J.; Bournel, A.; Dollfus, P. Wigner Monte Carlo simulation of phonon-induced electron decoherence in semiconductor nanodevices. *Phys. Rev. B* **2008**, *78*, 165306. [CrossRef]
15. Fischetti, M.V. Theory of electron transport in small semiconductor devices using the Pauli master equation. *J. Appl. Phys.* **1998**, *83*, 270–291. [CrossRef]
16. Fischetti, M.V. Master-equation approach to the study of electronic transport in small semiconductor devices. *Phys. Rev. B* **1999**, *59*, 4901. [CrossRef]
17. Oriols, X. Quantum-Trajectory Approach to Time-Dependent Transport in Mesoscopic Systems with Electron-Electron Interactions. *Phys. Rev. Lett.* **2007**, *98*, 066803. [CrossRef]
18. Marian, D.; Zanghì, N.; Oriols, X. Weak Values from Displacement Currents in Multiterminal Electron Devices. *Phys. Rev. Lett.* **2016**, *116*, 110404. [CrossRef]
19. Baym, G.; Kadanoff, L.P. Conservation Laws and Correlation Functions. *Phys. Rev.* **1961**, *124*, 287. [CrossRef]
20. Baym, G. Self-Consistent Approximations in Many-Body Systems. *Phys. Rev.* **1962**, *127*, 1391. [CrossRef]
21. Keldysh, L.V. Diagram Technique for Nonequilibrium Processes. *Sov. Phys. JETP (Zh. Eksp. Teor. Fiz.)* **1965**, *20*, 1018–1026.
22. Mahan, G.D. *Many-Particle Physics*; Plenum: New York, NY, USA, 1990.
23. Haug, H.; Jauho, A.P. *Quantum Kinetics in Transport and Optics of Semiconductors*; Vol. 123 of Springer Series in Solid-State Sciences; Springer: Berlin, Germany; New York, NY, USA, 1996.
24. Ferry, D.K.; Goodnick, S.M. *Transport in Nanostructures*; Cambridge University Press: Cambridge, UK, 1997.
25. Zhao, X.; Wei, C.M.; Yang, L.; Chou, M.Y. Quantum Confinement and Electronic Properties of Silicon Nanowires. *Phys. Rev. Lett.* **2004**, *92*, 236805. [CrossRef]
26. Pizzi, G.; Gibertini, M.; Dib, E.; Marzari, N.; Iannaccone, G.; Fiori, G. Performance of arsenene and antimonene double-gate MOSFETs from first principles. *Nat. Commun.* **2016**, *7*, 12585. [CrossRef]
27. Moussavou, M.; Cavassilas, N.; Dib, E.; Bescond, M. Influence of uniaxial strain in Si and Ge p-type double-gate metal-oxide-semiconductor field effect transistors. *J. Appl. Phys.* **2015**, *118*, 114503. [CrossRef]
28. Lherbier, A.; Persson, M.P.; Niquet, Y.M.; Triozon, F.; Roche, S. Quantum transport length scales in silicon-based semiconducting nanowires: Surface roughness effects. *Phys. Rev. B* **2008**, *77*, 085301. [CrossRef]
29. Luisier, M.; Klimeck, G. Atomistic full-band simulations of silicon nano-wire transistors: Effects of electron-phonon scattering. *Phys. Rev. B* **2009**, *80*, 155430. [CrossRef]
30. Mera, H.; Lannoo, M.; Li, C.; Cavassilas, N.; Bescond, M. Inelastic scattering in nanoscale devices: One-shot current-conserving lowest-order approximation. *Phys. Rev. B* **2012**, *86*, 161404. [CrossRef]
31. Mera, H.; Lannoo, M.; Cavassilas, N.; Bescond, M. Nanoscale device modeling using a conserving analytic continuation technique. *Phys. Rev. B* **2013**, *88*, 075147. [CrossRef]
32. Lee, Y.; Lannoo, M.; Cavassilas, N.; Luisier, M.; Bescond, M. Efficient quantum modeling of inelastic interactions in nanodevices. *Phys. Rev. B* **2016**, *93*, 205411. [CrossRef]
33. Lee, Y.; Bescond, M.; Cavassilas, N.; Logoteta, D.; Raymond, L.; Lannoo, M.; Luisier, M. Quantum treatment of phonon scattering for modeling of three-dimensional atomistic transport. *Phys. Rev. B* **2017**, *95*, 201412. [CrossRef]
34. Lee, Y.; Bescond, M.; Logoteta, D.; Cavassilas, N.; Lannoo, M.; Luisier, M. Anharmonic phonon-phonon scattering modeling of three-dimensional atomistic transport: An efficient quantum treatment. *Phys. Rev. B* **2018**, *97*, 205447. [CrossRef]
35. Caliceti, E.; Meyer-Hermann, M.; Ribeca, P.; Surzhykov, A.; Jentschura, U.D. From useful algorithms for slowly convergent series to physical predictions based on divergent perturbative expansions. *Phys. Rep.* **2007**, *446*, 1–96. [CrossRef]
36. Baker, G.A., Jr.; Graves-Morris, P. *Padé Approximants*; Cambridge University Press: Cambridge, UK, 1996.
37. Mera, H.; Pedersen, T.G.; Nikolić, B.K. Hypergeometric resummation of self-consistent sunset diagrams for steady-state electron-boson quantum many-body systems out of equilibrium. *Phys. Rev. B* **2016**, *94*, 165429. [CrossRef]
38. Hardy, G.H. *Divergent Series*; Chelsea: New York, NY, USA, 1991.

39. Svizhenko, A.; Anantram, M.P. Role of scattering in nanotransistors. *IEEE Trans. Electron Devices* **2003**, *50*, 1459. [CrossRef]
40. Jin, S.; Park, Y.J.; Min, H.S. A three-dimensional simulation of quantum transport in silicon nanowire transistor in the presence of electron-phonon interactions. *J. Appl. Phys.* **2006**, *99*, 123719. [CrossRef]
41. Carrillo-Nuñez, H.; Bescond, M.; Cavassilas, N.; Dib, E.; Lannoo, M. Influence of electron-phonon interactions in single dopant nanowire transistors. *J. Appl. Phys.* **2014**, *116*, 164505.
42. Luttinger, J.M.; Ward, J.C. Ground-State Energy of a Many-Fermion System. II. *Phys. Rev.* **1960**, *118*, 1417. [CrossRef]
43. Luttinger, J.M. Fermi Surface and Some Simple Equilibrium Properties of a System of Interacting Fermions. *Phys. Rev.* **1960**, *119*, 1153. [CrossRef]
44. Rhyner, R.; Luisier, M. Atomistic modeling of coupled electron-phonon transport in nanowire transistors. *Phys. Rev. B* **2014**, *89*, 235311. [CrossRef]
45. Shanks, D. Nonlinear transformations of divergent and slowly convergent sequences. *J. Math. Phys.* **1955**, *34*, 1–42. [CrossRef]
46. Bescond, M.; Li, C.; Mera, H.; Cavassilas, N.; Lannoo, M. Modeling inelastic pho-non scattering in atomic- and molecular-wire junctions. *J. Appl. Phys.* **2013**, *114*, 153712. [CrossRef]
47. Cavassilas, N.; Bescond, M.; Mera, H.; Lannoo, M. One-shot current conserving quantum transport modeling of phonon scattering in n-type double-gate field-effect-transistors. *Appl. Phys. Lett.* **2013**, *102*, 013508. [CrossRef]
48. Vajta, M. Some remarks on Pade-approximations. In Proceedings of the 3rd TEMPUS INTCOM Symposium on Intelligent Systems in Control and Measurement, Veszprém, Hungary, 9–14 September 2000; p. 53.
49. Fike, C.T. *Computer Evaluation of Mathematical Functions*; Prentice-Hall, Inc.: Englewood Cliffs, NJ, USA, 1968.
50. Bender, C.M.; Orszag, S.A. *Advanced Mathematical Methods for Scientists and Engineers I: Asymptotic Methods and Perturbation Theory*; Springer: New York, NY, USA, 1999.
51. Luisier, M. Atomistic modeling of anharmonic phonon-phonon scattering in nanowires. *Phys. Rev. B* **2012**, *86*, 245407. [CrossRef]
52. Luisier, M.; Schenk, A.; Fichtner, W. Atomistic simulation of nanowires in the sp3d5s* tight-binding formalism: From boundary conditions to strain calculations. *Phys. Rev. B* **2006**, *74*, 205323. [CrossRef]
53. Rhyner, R.; Luisier, M. Minimizing Self-Heating and Heat Dissipation in Ultrascaled Nanowire Transistors. *Nano Lett.* **2016**, *16*, 1022–1026. [CrossRef] [PubMed]
54. Slater, J.C.; Koster, G.F. Simplified LCAO Method for the Periodic Potential Problem. *Phys. Rev.* **1954**, *94*, 1498. [CrossRef]
55. Boykin, T.B.; Klimeck, G.; Oyafuso, F. Valence band effective-mass expressions in the sp3d5s* empirical tight-binding model applied to a Si and Ge parametrization. *Phys. Rev. B* **2004**, *69*, 115201. [CrossRef]
56. Sui, Z.; Herman, I.P. Effect of strain on phonons in Si, Ge, and Si/Ge heterostructures. *Phys. Rev. B* **1993**, *48*, 17938. [CrossRef] [PubMed]
57. Paul, A.; Luisier, M.; Klimeck, G. Modified valence force field approach for phonon dispersion: From zinc-blende bulk to nanowires. *J. Comput. Electron.* **2010**, *9*, 160. [CrossRef]
58. Valin, R.; Aldegunde, M.; Martinez, A.; Barker, J.R. Quantum transport of a nanowire field-effect transistor with complex phonon self-energy. *J. Appl. Phys.* **2014**, *116*, 084507. [CrossRef]
59. Cardona, M.; Ruf, T. Phonon self-energies in semiconductors: Anharmonic and isotopic contributions. *Solid State Commun.* **2001**, *117*, 201–212. [CrossRef]
60. Frey, M.; Esposito, A.; Schenk, A. Simulation of intravalley acoustic phonon scattering in silicon nanowires. In Proceedings of the 38th European Solid-State Device Research Conference (ESSDERC), Edinburgh, UK, 15–19 September 2008; p. 258.

MDPI
St. Alban-Anlage 66
4052 Basel
Switzerland
Tel. +41 61 683 77 34
Fax +41 61 302 89 18
www.mdpi.com

Materials Editorial Office
E-mail: materials@mdpi.com
www.mdpi.com/journal/materials

www.ingramcontent.com/pod-product-compliance
Lightning Source LLC
LaVergne TN
LVHW070614170726
843515LV00004B/75

* 9 7 8 3 0 3 9 3 6 2 0 8 0 *